Designing Hybrid Nanoparticles

Designing Hybrid Nanoparticles

Maria Benelmekki

Okinawa Institute of Science and Technology, Graduate University, Japan

Morgan & Claypool Publishers

ISBN   978-1-6270-5469-0 (ebook)
ISBN   978-1-6270-5468-3 (print)
ISBN   978-1-6270-5732-5 (mobi)

DOI   10.1088/978-1-6270-5469-0

Version: 20150401

IOP Concise Physics
ISSN 2053-2571 (online)
ISSN 2054-7307 (print)

A Morgan & Claypool publication as part of IOP Concise Physics
Published by Morgan & Claypool Publishers, 40 Oak Drive, San Rafael, CA, 94903, USA

IOP Publishing, Temple Circus, Temple Way, Bristol BS1 6HG, UK

To my family and many friends.

The ability to build something from the most fundamental constituents is a massive breakthrough.
Robert J Sawyer

Contents

Preface

The design of complex hybrid nanoparticles (HNPs) with advanced functionalities and the study of their fundamental properties play a major role in the development of a new generation of nanostructured materials. The possibility of tailoring the dimensions, composition and structure of HNPs represents a major milestone in the control of their unique physico-chemical properties. These properties, combined with the ability to produce high-quality HNPs in fairly large numbers for a reasonable cost, led to their potential applicability in fields ranging from materials science to nanomedicine. Within the field of nanomedicine, the amazing properties of HNPs and the possibility of integrating diagnostic and therapeutic entities onto a single nanoparticle (NP) have made them an interesting tool in many applications such as multimodal bioimaging and targeted drug delivery.

In the last few years, several 'bottom-up' and 'top-down' synthesis routes have been developed to produce tailored HNPs. This book provides new insight into one of the most promising 'bottom-up' techniques, based on a practical magnetron-sputtering-based inert-gas-condensation (MS-IGC) method. A modified MS-IGC system is presented and its performance under different conditions is evaluated.

This book is designed for graduate students, process engineers, and researchers in physics, materials science, biophysics and related fields. It meets the critical need of furthering understanding of the fundamentals behind the design and tailoring of the NPs produced by the MS-IGC method. It shows that the morphology, size and properties of NPs can be modulated by tuning the deposition parameters, such as the energy, cooling rate, and collision and coalescence processes, experienced by the NPs during their formation. The formation mechanisms of different HNPs are suggested, combining the physico-chemical properties of the materials with the experimental conditions.

This book illustrates the potential of the MS-IGC method in synthesizing multifunctional NPs and nanocomposites with accurate control of their morphology and structure. However, for a better understanding of HNP formation, further improvements in the characterization methods for aggregation zone conditions are needed. In addition, the optimization of the yield and harvesting process of HNPs is essential to make this method sufficiently attractive for large-scale production.

Acknowledgments

I would like to thank Dr E Xuriguera, Dr R E Diaz, Mr T Sasaki and Mr A Bright for their technical support with TEM; Mr K Baughman and Dr J Vernieres for their technical support with SEM; Dr A Roberts for his introduction to the Kratos Ultra system; and Dr Ll M Martinez for his technical support and useful commentary in using the FEM software package for magnetic field mapping.

Author biography

Maria Benelmekki

Maria Benelmekki completed a PhD in experimental solid state physics (1997) at the Autonomous University of Barcelona (UAB-Spain). After postdoctoral work at Ecole National Supérieure d'Art et Métier (ENSAM-France), she entered industrial research as a project manager, working in research and development in international projects related to a broad range of nanomaterials and their applications in the food packaging and automotive industries. Eight years later, she joined the Centre of Physics of Minho University (Portugal) where she focused her research on the field of nanoparticles and hybrid nanomaterials. She extended her research to the fundamental aspects of material surfaces and interfaces and their performance for biomagnetic separation. She is currently a senior staff scientist at OIST-Graduate University, Japan.

Technical acronyms

MS-IGC	magnetron-sputtering-based inert-gas-condensation
NPs	nanoparticles
HNPs	hybrid nanoparticles
MPNPs	magneto-plasmonic nanoparticles
TEM	transmission electron microscopy
HRTEM	high-resolution transmission electron microscopy
STEM	scanning transmission electron microscopy
XPS	x-ray photoelectron spectroscopy
SEM	scanning electron spectroscopy
EELS	electron energy loss spectroscopy
PDF	probability distribution function
FFT	fast Fourier transform
CDF	cumulative distribution function

Chapter 1

An introduction to nanoparticles and nanotechnology

1.1 An overview of nanoparticles and nanotechnologies

'Nano' is a prefix used to describe 'one billionth', or 10^{-9}, of something. The concept of nanotechnology was introduced by physics Nobel laureate Richard P Feynman in his famous lecture entitled 'There's plenty of room at the bottom' at the December 1959 meeting of the American Physical Society [1]. Since then, there have been many revolutionary developments in physics, chemistry and biology that have demonstrated Feynman's ideas of manipulating matter at the atomic scale. In 1974, Norio Taniguchi (a professor at the Tokyo University of Science) invented the term 'nanotechnology' to describe extra-high precision and ultra-fine dimensions. He introduced the 'top-down approach' by predicting improvements and miniaturization in integrated circuits, optoelectronic devices, mechanical devices and computer memory devices. Approximately ten years later, K Eric Drexler introduced the 'bottom-up approach' when he discussed the creation of larger objects from their atomic and molecular components as the future of nanotechnology [2].

Nanotechnologies are now widely considered to have the potential to bring benefits in areas as diverse as drug development, water decontamination, information and communication technologies, and the production of stronger and lighter materials. Nanotechnologies involve the creation and manipulation of materials at the nanometre scale, either by scaling up from single groups of atoms or by refining or reducing bulk materials [3].

While the development of nanotechnologies is a modern multidisciplinary science involving the fields of physics, chemistry, biology and engineering, the production of nanoparticles (NPs), both in nature and by humans, dates from the pre-Christian era. For example, the Romans introduced metals with nanometric dimensions in glass-making; the famous Lycurgus cup (currently exhibited at the British Museum), which displays a different colour depending on whether it is illuminated externally

doi:10.1088/978-1-6270-5469-0ch1

Table 1.1. A non-exhaustive list of nanomaterials, either used in industry or under investigation [14].

aluminium	dendrimers	platinum
aluminium oxide	dimethyl siloxide	polyethylene
aluminium hydroxide	dysprosium oxide	polystyrene
antimony oxide	fullerenes	praseodymium oxide
antimony pentoxide	germanium oxide	rhodium
barium carbonate	indium oxide	samarium oxide
bismuth oxide	iron and iron oxides	silanamine
boron oxide	lanthanum oxide	silicon dioxide
calcium oxide	lithium titanate	silver
carbon black	manganese oxide	carbon nanotubes
cerium oxide	molybdenum oxide	tantalum
chromium oxide	nanoclays	terbium oxide
cluster diamonds	neodymium oxide	titanium dioxide
cobalt and cobalt oxide	nickel	tungsten
colloidal gold	niobium	yttrium oxide
copper (II) oxide	palladium	zinc oxide

(green) or internally (red), contains NPs of silver and gold [4]. In 1857, Faraday reported the synthesis of colloidal gold (and other metals such as Cu, Zn, Fe and Sn) and its interaction with light [5]. For an overview and chronological table of nanotechnologies, see [6].

Another example of interest is the case of magnetic NPs. Krishnan [7], illustrated the role that magnetic materials play in biology and medicine. In the field of magnetic NPs, a noteworthy pioneering work was published by Blakemore in 1975 [8], where biochemically precipitated magnetite (Fe_3O_4) was found in the tissues of various organisms including bacteria, algae, insects, birds and mammals. Many of these organisms use biogenic magnetite to sense the Earth's magnetic field for orientation and navigation. For more details on the development of magnetic NP synthesis and its presence in biomedicine and biotechnology see [7].

Throughout the last century, the field of colloid science has developed enormously and has been used to produce many materials, including metals, oxides and organic products [9, 10]. One of the first and most easily prepared magnetic colloidal systems was developed by Stephen Papell of the National Aeronautics and Space Administration in the early 1960s [11]. Papell's colloid consisted of finely divided particles of magnetite suspended in paraffin. To prevent particle–particle agglomeration or sedimentation, Papell added oleic acid as a dispersing agent. Subsequently, similar magnetic suspensions have also been synthesized with different nanometre sized particles of pure elements, such as iron, nickel and cobalt, in a wide range of carrier liquids [12, 13].

Ordinary materials, when reduced to the nanoscale, often exhibit novel and unpredictable characteristics such as extraordinary strength, chemical reactivity, electrical conductivity, superparamagnetic behaviour and other characteristics that the same material does not possess at the micro- or macroscale. A huge range of

Table 1.2. General classification and potential applications of NPs [14].

Product areas with end-products containing NPs	Sectors where nanotechnology is expected to have a considerable impact
• Cosmetics and personal care products • Paints and coatings • Household products • Catalysts and lubricants • Sports products and textiles • Medical and healthcare products • Food and nutritional ingredients • Food packaging and agrochemicals • Veterinary medicines • Construction materials • Consumer electronics	• Medical and pharmaceutical sector • Bio-nanotechnology, bio-sensors • Energy sector, including fuel cells, batteries and photovoltaics • Environment sector including water remediation • Automotive sector • Aeronautics sector • Construction sector, including reinforcement of materials • Composite materials • Electronics and optoelectronics, photonics

nanomaterials is currently being produced at an industrial scale, while others are being produced at smaller scales as they are still under research and development (table 1.1).

In summary, a number of examples of NPs with new magnetic, catalytic, magneto-optical or optical properties, among others, that differ from those of the bulk materials have been reported in the scientific literature. Size reduction has been found to be the reason behind many of these novel physical and chemical properties, which allow a wide range of applications with economic benefits. In 2010, more than 1000 products containing NPs became commercially available (table 1.2.) [15, 16].

1.2 Classification of nanomaterials

Typically, NPs are defined as an agglomeration of atoms and molecules in the range of 1–100 nm. They can be composed of one or more species of atoms (or molecules) and can exhibit a wide range of size-dependent properties. Within this size range, NPs bridge the gap between small molecules and bulk materials in terms of energy states [17]. NPs are generally classified based on their dimensionality, morphology, composition, uniformity and agglomeration [18, 19].

1.2.1 Dimensionality

1D nanomaterials. Materials with one dimension in the nanometre scale are typically thin films or surface coatings. Thin films have been developed and used for decades in various fields including electronics, information storage systems, chemical and biological sensors, fibre-optic systems, and magneto-optic and optical devices. Thin films can be deposited by various methods and can be grown controllably at the atomic level (a monolayer) [20].

2D nanomaterials. 2D nanomaterials have two dimensions in the nanometre scale. These include for example, nanotubes, dendrimers, nanowires, fibres and fibrils.

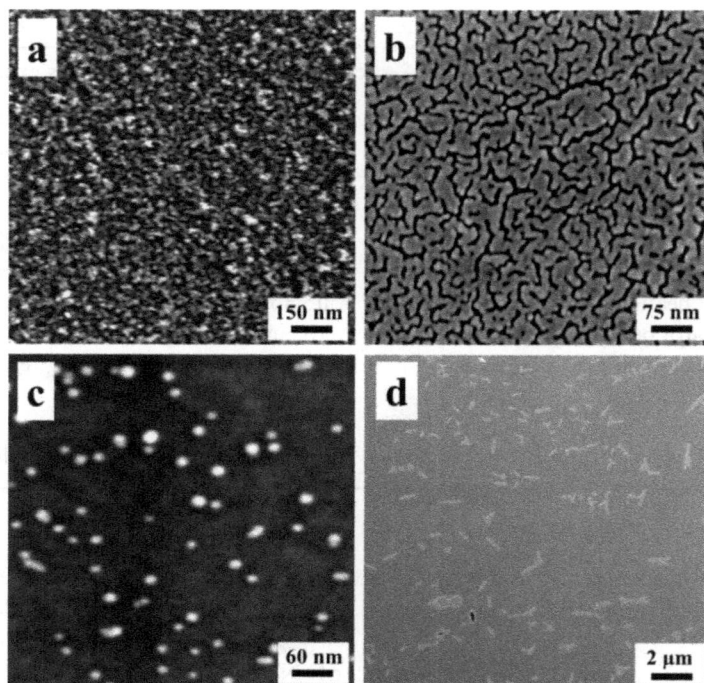

Figure 1.1. Scanning electron microscopy (SEM) images showing (*a*) a film of Ti NPs of 80 nm thickness, (*b*) a near-percolating Au film, (*c*) monodispersed Cu NPs and (*d*) Fe nanorods.

Free particles with a large aspect ratio with dimensions in the nanoscale range are also considered to be 2D nanomaterials. The properties of 2D systems are less well understood and their manufacturing capabilities are less advanced.

3D nanomaterials. Materials that are nanoscale in all three dimensions are considered to be 3D nanomaterials. These include quantum dots or nanocrystals, fullerenes, particles, precipitates and colloids. Some 3D systems, such as natural nanomaterials and combustion products, metallic oxides, carbon black, titanium oxide (TiO_2) and zinc oxide (ZnO) are well known, while others such as fullerenes, dendrimers and quantum dots represent the greatest challenges in terms of production and understanding of properties.

Figure 1.1 shows examples of nanomaterials with different dimensions. All the samples were deposited on a Si (111) substrate using the magnetron-sputtering-based inert-gas-condensation (MS-IGC) method as described in figures 2.1 and 2.2. The materials shown in figures 1.1(*a*) and (*b*) can be classified as 1D nanomaterials, while the Cu NPs shown in figure 1.1(*c*) are classified as 3D nanomaterials. The iron nanorods shown in figure 1.1(*d*) can be classified as 2D nanomaterials.

1.2.2 The morphology of NPs and nanocomposites

The morphological characteristics to be taken into account are the flatness, aspect ratio and spatial position of each element in the case of hybrid NPs (HNPs).

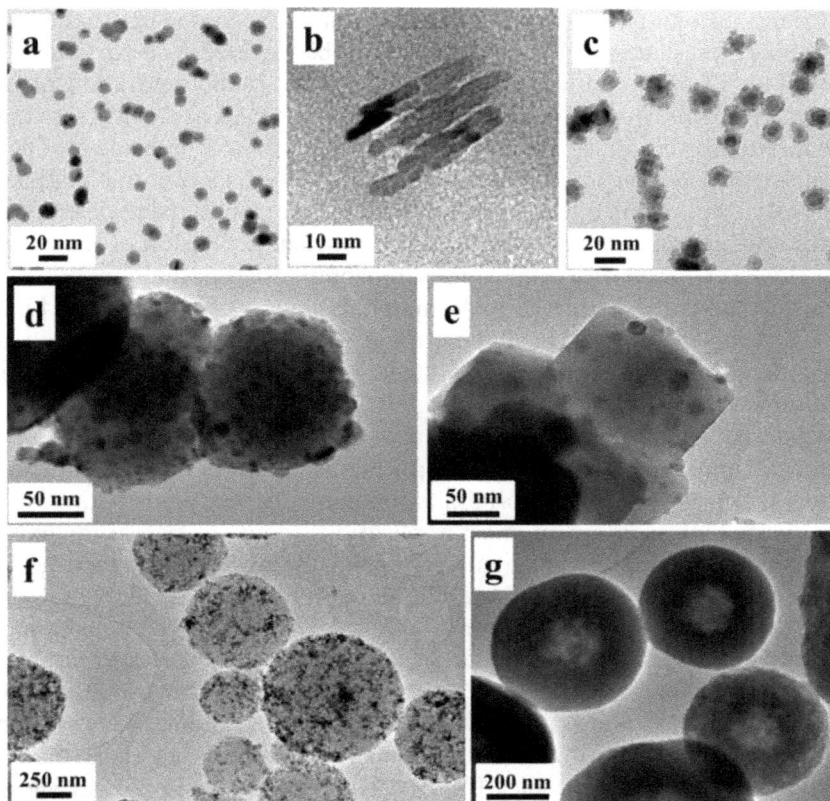

Figure 1.2. TEM images of examples of NPs with different morphologies and compositions. (*a*) Monodispersed Cu NPs, (*b*) Fe nanorods, (*c*) Cu–Si core–shell NPs, (*d*) porous Fe_3O_4 NPs, (*e*) Fe_3O_4 cubes decorated with Ni NPs, (*f*) porous silica spheres with γ-Fe_2O_3 NPs adsorbed on their surfaces and (*g*) γ-Fe_2O_3 NPs embedded in porous silica spheres. For more details about the preparation and characterization of these composites see [22, 23].

A general classification exists between high and low aspect ratio particles. High aspect ratio NPs include nanotubes and nanowires. Small aspect ratio morphologies include spherical, oval, cubic, prism, helical and pillar shapes. Figure 1.2 shows examples of different morphologies of NPs and nanocomposites. Transmission electron microscopy (TEM) images of monodispersed Cu NPs, Fe nanorods and Cu core–Si shell NPs are shown in figures 1.2(*a*), (*b*) and (*c*), respectively. The details of the preparation methods for these NPs are presented in chapter 2. The TEM images in figures 1.2(*d*) and (*e*) show a porous magnetite NP and magnetite cubes decorated with Ni nanocrystals, respectively. These NPs were designed and synthesized using the hydrothermal process for purification of histidine-tagged proteins [21].

With regard to nanocomposites, substantial progress has been made in recent years in developing technologies in the fields of magnetic microspheres, magnetic nanospheres and ferrofluids. Nanospheres and microspheres containing a magnetic core embedded in a non-magnetic matrix are used in numerous biological applications [7]. They are used, for example, as carriers that can be targeted to a particular site by using an external magnetic field. In addition, the magnetic separation of organic compounds, proteins, nucleic acids and other biomolecules and cells from complex reaction mixtures is becoming the most suitable method for large scale production in bioindustrial purification and extraction processes. For *in vivo* applications, it is imperative that well-defined biocompatible coatings surround the magnetic particles to prevent any aggregation and also to enable efficient protection of the body from toxicity. However, for *in vitro* applications, biocompatible coatings are not essential; particles can be coated with non-toxic materials inert to chemical and biological media. The particles employed in all these applications are mainly superparamagnetic colloids with appropriate coatings, guaranteeing the stability and biocompatibility of the solutions.

Superparamagnetic NPs exhibit magnetizations of magnitudes similar to those of ferromagnetic materials, however, they have neither coercivity nor remanence. This behaviour, which is of quantum origin, is limited to nanocrystals with sizes below the critical size [24]. Conversely, most applications require superparamagnetic colloidal dispersions with large magnetic responses. Because the magnetization of a particle is proportional to its volume, the maximum magnetization that one can achieve is limited by the critical size of the superparamagnetic transition, which depends on the material [7]. A well-established strategy to create superparamagnetic particles with larger superparamagnetic responses is using nanocomposites (see, for example, figures 1.2(*f*) and (*g*)). These superparamagnetic composites are typically made by embedding superparamagnetic nanocrystals in a non-magnetic matrix such as polystyrene or nanoporous silica [25–27]. The resulting colloidal particles retain the superparamagnetic response of their constituent nanocrystals and show larger magnetization when an external magnetic field is applied. Furthermore, neither coercivity nor remanence are observed at the working temperature. However, in addition to the intrinsic superparamagnetic behaviour of the constituent NPs, one must consider the interactions between the NPs inside the skeleton matrix due to their proximity and surface effects due to the coating; these can lead to changes in the overall magnetic response of the colloidal particle.

1.2.3 NP chemical composition

NPs can be composed of a single constituent material or be a composite of several materials. The NPs found in nature are often agglomerations of materials with various compositions, while pure single-composition materials can be easily synthesized using a variety of methods (see chapter 2).

There are three main types of chemical ordering in HNPs (figure 1.3) that describe the way in which the atoms of the elements are arranged within the same NP [28, 29]:

- *Mixed NPs* can be either random or ordered (figure 1.3(*a*)). Randomly mixed alloys correspond to solutions of solids, whereas ordered nanoalloys correspond to ordered arrangements of A and B atoms.
- *Core–shell NPs* consist of a shell of one type of atom (B) surrounding a core of another type of atom (A) (figure 1.3(*b*)). This pattern is generally denoted

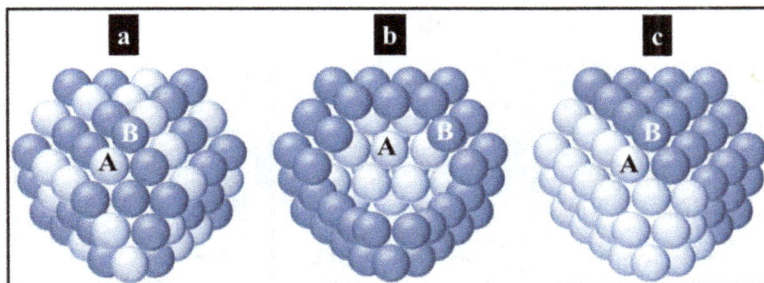

Figure 1.3. Schematic images of binary NPs: a mixed structure (*a*), a core–shell structure (*b*) and a layered structure (*c*) of A and B elements.

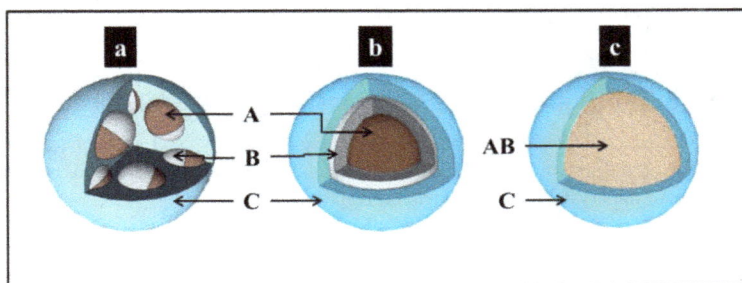

Figure 1.4. Schematic images of ternary NPs formed of elements A, B and C: (*a*) a multicore–shell morphology (the cores present a dumbbell-like morphology), (*b*) a core–multishell morphology and (*c*) an alloyed-core–shell morphology.

by A@B and is common for a large class of NPs. Various thermodynamic considerations, discussed further in chapters 3 and 4, lead to the segregation of materials within the core or shell. A subset of the core–shell category consists of multishell (or 'onion-like') NPs. These NPs have alternating A–B–A shells, or A–B–C in the case of ternary NPs as depicted in figure 1.4(*b*). The formation of these latter structures is discussed in chapter 4, where it is demonstrated that Fe–Ag–Si multishell structures are obtained by adjusting the experimental conditions.

- *Layered NPs* are commonly referred to in the literature as Janus (or 'dumbbell-like') NPs. They consist of two types of NPs (A and B) sharing a common interface (figure 1.3(*c*)). These types of NPs tend to minimize the number of bonds between elements A and B. This heterojunction structure facilitates phase separation.

Because of the increasing need for multifunctional NPs, other complex structures of NPs such as multicore–shell structures in which the cores can present either 'dumbbell-like' or 'onion-like' structures have been reported in the literature [30–34]. This type of NP is presented and discussed in chapter 4. Another multifunctional

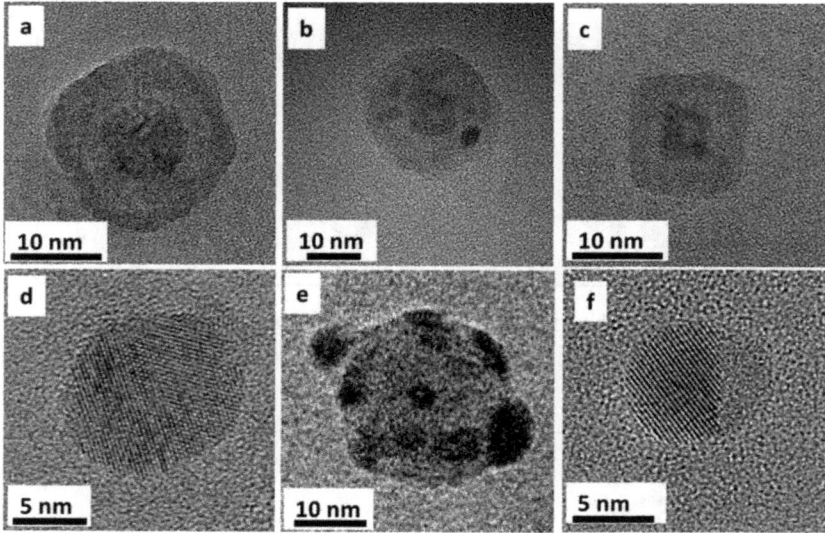

Figure 1.5. (*a*) A Cu@Ag core–shell NP, (*b*) a Cu@Si multicore–shell NP, (*c*) a Fe@Fe$_2$O$_3$ core–shell NP, (*d*) a CuAg mixed NP, (*e*) a Si NP inoculated with Ag NPs resulting in a satellite morphology and (*f*) a FeAg dumbbell-like NP where the crystalline hemisphere corresponds to Ag.

subset of the core–shell arrangement is an alloyed single core NP encapsulated in an inert shell (figure 1.4(*c*)). The case of FeAl@Al$_2$O$_3$ NPs [35] is presented and discussed in section 3.2.

TEM images of NPs with various compositions and morphologies are shown in figure 1.5. These NPs were prepared using the MS-IGC method. Control over the composition and morphology of these NPs was achieved by adjusting the experimental conditions as explained in chapters 3 and 4. A detailed description of the deposition system is given in chapter 2.

1.3 NP uniformity and agglomeration

Based on their chemistry and electromagnetic properties, NPs can exist as dispersed aerosols, suspensions/colloids or in an agglomerate state. In fact, NPs tend to adhere to each other and to form agglomerates because of van der Waals forces that act over short distances, magnetic interactions, electrostatic forces present in the particles and adhesion forces related to the liquids adsorbed on their surfaces. Agglomeration due to Brownian motion is classified as 'coagulation'.

To avoid agglomeration, several processes include a post-synthesis stage to modify the particle surface by coating it with another organic or inorganic substance. In an agglomerate state, NPs may behave as larger particles, depending on the size of the agglomerate. For example, magnetic NPs tend to cluster, forming an agglomerate state, unless their surfaces are coated with a non-magnetic material. Figure 1.6 illustrates the typical process of stabilization of γ-Fe$_2$O$_3$ NPs in an aqueous suspension. The molecules of the anionic surfactant sodium dodecyl

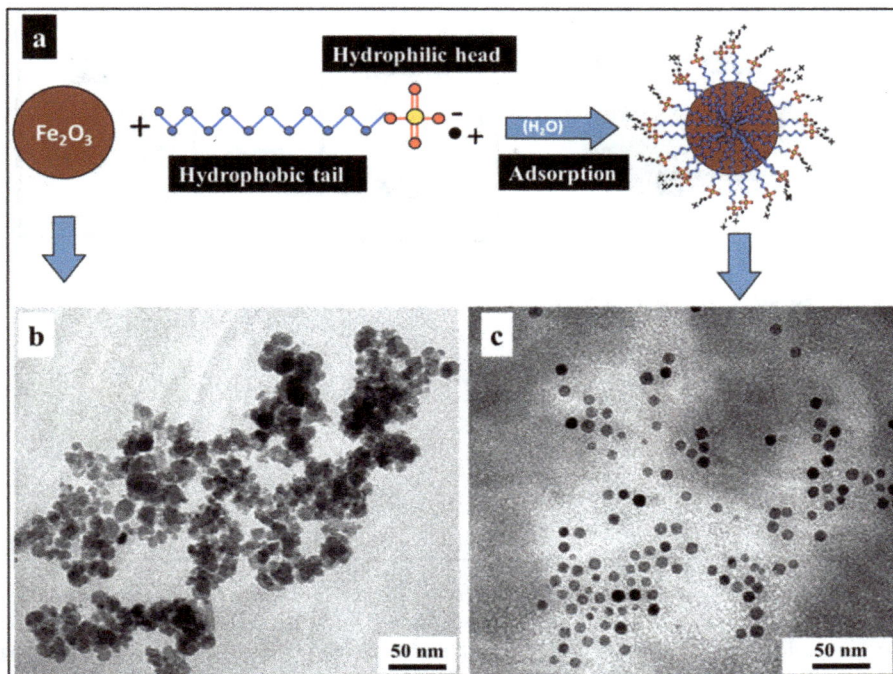

Figure 1.6. (*a*) An iIllustration of the stabilization process applied to γ-Fe$_2$O$_3$ NPs using SDS surfactant. (*b*) A TEM image of the precipitated Fe$_2$O$_3$ NPs without SDS. (*c*) A TEM image of SDS-modified Fe$_2$O$_3$ NPs (adapted from [36] by permission of The Royal Society of Chemistry).

sulphate (SDS) are adsorbed onto the surfaces of the nanocrystals providing a negative charge in water. Therefore, the nanocrystals in the solution repel each other electrostatically resulting in a stable colloidal suspension. For more details on this procedure see [36].

In the case of NPs deposited using vapour phase methods, the NPs are generally deposited on a solid substrate. The transfer of these NPs to a stable suspension is still under investigation. For example, attempts were made to co-deposit NPs from the vapour phase with a beam of water vapour, methanol, or isopropanol onto a nitrogen-cooled substrate [37, 38]. However, the stability of the resulting suspensions was not reported. Recently, a simple and environmentally friendly method for harvesting NPs was developed using polyvinylpyrrolidone (PVP) as a stabilizer [32]. PVP was selected as a non-toxic polymer with good wetting properties. Figure 1.7 shows the procedure used to harvest the NPs to a stable and homogeneous colloidal suspension.

1.4 NP characterization

Once NPs are synthesized, it is important to fully characterize and understand their structure. Over the years, many methods have been developed for this purpose. In this section, the focus will be on the main techniques with relevance to this book,

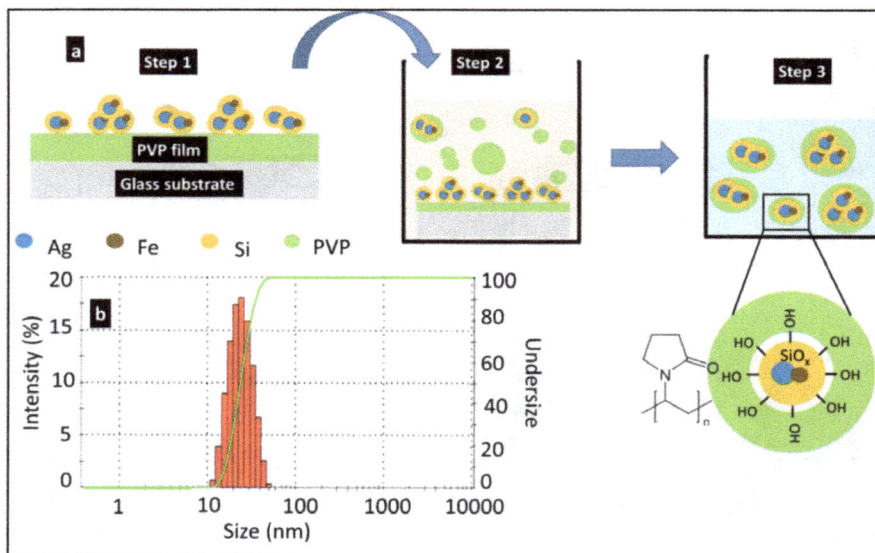

Figure 1.7. (*a*) A schematic of the exfoliation procedure for the NPs. Step 1: multicore–shell NPs were deposited on a spin-coated PVP film on a glass substrate. Step 2: the NP/PVP/glass samples were immersed in methanol and sonicated for 15 minutes and then separated to remove excess PVP. Step 3: after washing the precipitated NPs with methanol, they were re-suspended in ultrapure water. (*b*) A dynamic light scattering histogram showing the size distribution of the HNPs. Reproduced from [32] by permission of The Royal Society of Chemistry.

namely TEM, scanning TEM (STEM), electron energy loss spectroscopy (EELS) and x-ray photoelectron spectroscopy (XPS).

TEM is a very powerful technique for the characterization of NP size, composition and crystalline structure. When an electron beam interacts with a sample, the electrons can be either transmitted, scattered, backscattered or diffracted [39, 40]. TEM uses the transmitted electron signal to form an image of the sample. The transmitted electron beam is dependent on the sample thickness; for thin samples (a few nanometres), the transmitted electrons pass through without significant energy loss. STEM differs from TEM by focusing the electron beam into a narrow spot that is scanned over the sample in a raster.

Because the attenuation of the electrons depends significantly on the density and thickness of the sample, the transmitted electron beam forms a 2D image of the sample. In hybrid samples, STEM imaging allows the identification of different components based on intensity variation. This intensity variation is related to the difference in the atomic numbers of each component (Z-contrast). In addition, the rastering of the beam across the sample makes it possible to couple STEM with other characterization methods such as EELS [41], allowing direct correlation of image and quantitative data thus obtain details regarding the chemical composition of NPs.

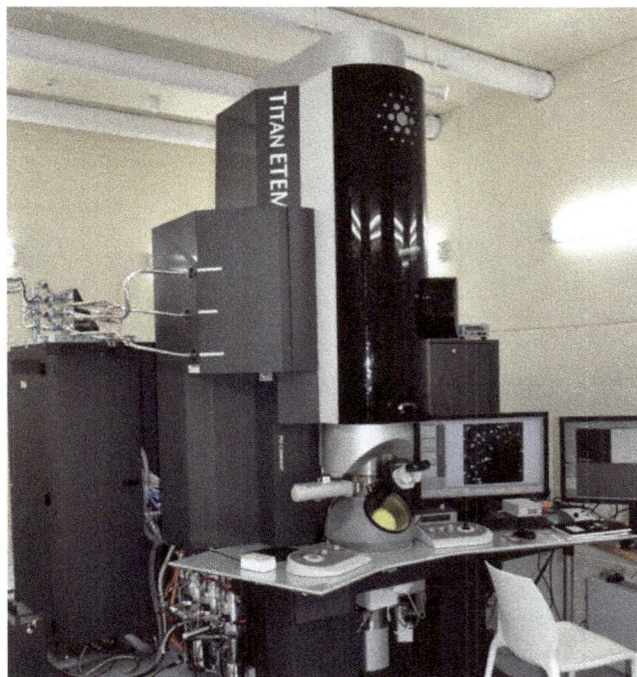

Figure 1.8. The transmission electron microscope used for the characterization of the NPs presented in this book. (Credit: OIST-Graduate University.)

General TEM analysis does not have sufficient resolution to determine the crystallinity of a nanomaterial. However, high-resolution TEM (HRTEM) can be successfully employed for the characterization of the crystallinity of a sample with atomic resolution, as well as for providing information regarding electron diffraction analysis. This approach helps in gaining insight into the ordering of atoms in a NP. Figure 1.8 shows a photograph of the transmission electron microscope used in the characterization of the NPs presented in this book.

XPS is a surface-sensitive quantitative spectroscopic technique that measures the elemental composition, chemical states and electronic states of the elements within the material. XPS spectra are obtained by irradiating a material with a beam of x-rays while simultaneously measuring the kinetic energy of the electrons that escape from the top 0 to 10 nm of the material being analysed. XPS requires high vacuum ($P \sim 10^{-8}$ mbar) or ultra-high vacuum ($P < 10^{-9}$ mbar) conditions. However, when used to analyse NPs, the importance of the coverage of the NPs must be kept in mind: high coverage leads to high-quality spectra. On the other hand, to quantify the composition of the NPs, an inert transfer of the sample to the analysis chamber is necessary to avoid contamination and oxidation of the NP surface. The system used to analyse the NPs presented in this book is shown in figure 1.9.

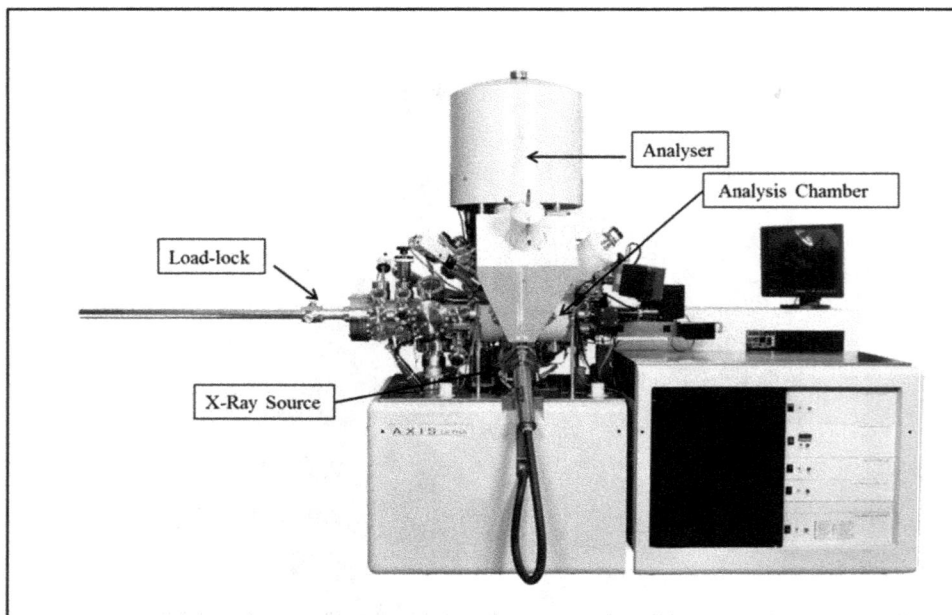

Figure 1.9. The XPS system used to analyse the NPs presented in chapters 3 and 4 of this book [42].

References

[1] Feynman R P 1960 There's plenty of room at the bottom *Engineering and Science* **23** 22–36
[2] Drexler K E 1990 *Engines of Creation: The Coming Era of Nanotechnology* (Oxford: Oxford University Press)
[3] National Nanotechnology Initiative www.nano.gov
[4] The Lycurgus Cup www.britishmuseum.org/explore/highlights/highlight_objects/pe_mla/t/the_lycurgus_cup.aspx
[5] Faraday M 1857 The Bakerian lecture: experimental relations of gold (and other metals) to light *Phil. Trans. R. Soc. Lond.* **147** 145–81
[6] Satoshi Horikoshi S and Serpone N (ed) 2013 *Microwaves in Nanoparticle Synthesis* 1st edn (Berlin: Wiley)
[7] Krishnan K M 2010 Biomedical nanomagnetics: a spin through possibilities in imaging diagnostics and therapy *IEEE Trans. Magn.* **46** 2523–58
[8] Blakemore R 1975 Magnetotactic bacteria *Science* **190** 377–9
[9] Aitken R J, Creely K S and Tran C L 2004 Nanoparticles: an occupational hygiene review *Health and Safety Executive Report* www.hse.gov.uk/research/rrpdf/rr274.pdf
[10] Ostiguy C, Lapointe G, Ménard L, Cloutier Y, Trottier M, Boutin M, Antoun M and Normand C 2006 Nanoparticles: actual knowledge about occupational health and safety risks and prevention measures *IRSST Report* www.irsst.qc.ca/files/documents/pubirsst/r-470.pdf
[11] Papell S 1963 Low viscosity magnetic fluid obtained by the colloidal suspension of magnetic particles *US Patent Specification* 3215572A
[12] Rosensweig R E 1982 Magnetic fluids *Sci. Am.* **10** 136–45

[13] Kaiser R and Miskolczy G 1970 Magnetic properties of stable dispersions of subdomain magnetite particles *J. Appl. Phys.* **41** 1064–72

[14] Lövestam G, Rauscher H, Roebben G, Sokull Klüttgen B, Gibson N, Putaud J-P and Stamm H 2010 Considerations on a definition of nanomaterial for regulatory purposes *JRC Reference Report* https://ec.europa.eu/jrc/sites/default/files/jrc_reference_report_201007_nanomaterials.pdf

[15] Ostiguy C, Roberge B, Woods C and Soucy B 2010 Engineered nanoparticles: current knowledge about OHS risks and prevention measures *IRSST Studies and Research Projects* report 656, 2nd edn

[16] Sebastian V, Arruebo M and Santamaria J 2014 Reaction engineering strategies for the production of inorganic nanomaterials *Small* **10** 835–53

[17] Johnston R L and Wilcoxon J P (ed) 2012 *Frontiers of Nanoscience* (Oxford: Elsevier)

[18] Royal Academy of Engineering and Royal Society 2004 *Nanoscience and Nanotechnologies: Opportunities and Uncertainties* http://www.nanotec.org.uk/finalReport.htm

[19] Buzea C, Pacheco-Blandino I and Robbie K 2007 Nanomaterials and nanoparticles: sources and toxicity *Biointerphases* **2** MR17–MR172

[20] Seshan K (ed) 2002 *Handbook of Thin-Film Deposition Processes and Techniques— Principles, Methods, Equipment and Applications* (Norwich, NY: William Andrew/Noyes)

[21] Benelmekki M, Xuriguera E, Caparros C, Corchero J L and Lanceros-Mendez S 2012 Nonionic surfactant assisted hydrothermal growth of porous iron oxide nano-spheres and its application for *His*-tagged proteins capturation, unpublished results

[22] Benelmekki M, Xuriguera E, Caparros C, Rodriguez E, Mendoza R, Corchero J L, Lanceros-Mendez S and Martinez L M 2012 Design and characterization of Ni^{2+} and Co^{2+} decorated porous magnetic silica spheres synthesised by hydrothermal assisted modified-Stöber method for *His*-tagged proteins separation *J. Colloid Interface Sci.* **365** 156–62

[23] Caparros C, Benelmekki M, Martins P, Xuriguera E, Ribeiro C J, Martinez L M and Lanceros-Mendez S 2012 Hydrothermal assisted synthesis of iron oxide-based magnetic silica spheres and their performance in magnetophoretic water purification *Mater. Chem. Phys.* **135** 510–7

[24] Bean C P and Livingstone J D 1959 Superparamagnetism *J. Appl. Phys.* **30** 120–9

[25] Leun D and Sengupta A K 2000 Preparation and characterization of magnetically active polymeric particles (MAPPs) for complex environmental separations *Environ. Sci. Technol.* **34** 3276–82

[26] Jain T K, Richey J, Strand M, Leslie-Pelecky D L, Flask C A and Labhasetwar V 2008 Magnetic nanoparticles with dual functional properties: drug delivery and magnetic resonance imaging *Biomaterials* **29** 4012–21

[27] Behrens S 2011 Preparation of functional magnetic nanocomposites and hybrid materials: recent progress and future directions *Nanoscale* **3** 877–92

[28] Paz-Borbon L O 2011 *Computational Analysis of Transition Metal Nanoalloys* (Berlin: Springer)

[29] Tiruvalam R C, Pritchard J C, Dimitratos N, Lopez-Sanchez J A, Edwards J K, Carley A F, Hutchins G J and Kiely C J 2011 Aberration corrected analytical electron microscopy studies of sol-immobilized Au + Pd, Au{Pd} and Pd{Au} catalysts used for benzyl alcohol oxidation and hydrogen peroxide production *Faraday Discuss.* **152** 63–86

[30] Shi W, Zeng H, Sahoo Y, Ohulchanskyy T Y, Ding Y, Wang Z L, Swihart M and Prasad P N 2006 A General Approach to Binary and Ternary Hybrid Nanocrystals *Nano Lett.* **6** 875–81

[31] Llamosa Perez D, Espinosa A, Martinez L, Roman E, Ballesteros C, Mayoral A, Garcia-Hernandez M and Huttel Y 2013 Thermal diffusion at nanoscale: from CoAu alloy nanoparticles to Co@Au core/shell structures *J. Phys. Chem. C.* **117** 3101–8

[32] Benelmekki M, Bohra M, Kim J H, Diaz R E, Vernieres J, Grammatikopoulos P and Sowwan M 2014 Facile Single-Step Synthesis of Ternary Multicore Magneto-Plasmonic Nanoparticles *Nanoscale* **6** 3532–5

[33] Benelmekki M, Vernieres J, Kim J H, Diaz R E, Grammatikopoulos P and Sowwan M 2015 On the Formation of Ternary Metallic–Dielectric Multicore–Shell Nanoparticles by Inert-Gas Condensation Method *Mater. Chem. Phys.* **151** 275–81

[34] Benelmekki M and Sowwan M 2015 *Handbook of Nano-Ceramic and Nano-Composite Coatings and Materials* ed H Makhlouf and D Scharnweber (Amsterdam: Elsevier) at press, ch 1

[35] Vernieres J, Benelmekki M, Kim J H, Grammatikopoulos P, Bobo J F, Diaz R E and Sowwan M 2014 Single-step gas phase synthesis of stable iron aluminide nanoparticles with soft magnetic properties *APL MATERIALS* **2** 116105

[36] Benelmekki M, Martinez L M, Andreu J S, Camacho J and Faraudo J 2012 Magnetophoresis of colloidal particles in a dispersion of superparamagnetic nanoparticles: Theory and Experiments *Soft Matter* **8** 6039–47

[37] Haeften K V, Binns C, Brewer A, Crisan O, Howes P B, Lowe M P, Sibbley-Allen C and Thornton S C 2009 A novel approach towards the production of luminescent silicon nanoparticles: sputtering, gas aggregation and co-deposition with H_2O *Eur. Phys. J. D* **52** 11–14

[38] Galinis G, Yazdanfar H, Bayliss M, Watkins M and Haeften K V 2012 Towards biosensing via fluorescent surface site of nanoparticles *J. Nanopart. Res.* **14** 1019

[39] Williams D B and Carter B C 2009 *Transmission Electron Microscopy: A Text Book for Material Science* 2nd edn (Berlin: Springer)

[40] Frenkel A I, Hills C W and Nuzzo R G 2001 A View from the Inside: Complexity in the Atomic Scale Ordering of Supported Metal Nanoparticles *J. Phys. Chem. B* **105** 12689

[41] Pearson D H, Ahn C C and Fultz B 1993 White lines and d-electron occupancies for the 3rd and 4th transition metals *Phys. Rev. B* **47** 8471–8

[42] Kratos Analytical www.kratos.com

Chapter 2

Production of hybrid nanoparticles

2.1 An overview of the production methods for NPs

Different methods of production are used to optimize specific properties of NPs, the yield and the suitability for scaling up. Broadly speaking, methods of NP production can be divided in two main groups: 1) top-down methods and 2) bottom-up methods. Other emerging and promising lithographic procedures, such as e-beam and ion-beam nanolithography and scanning-probe-based lithography, are of high interest in the lab because of the excellent control afforded at the sub-100 nm scale. However, these procedures are not practical for large-scale production due to their high cost and low yields. Nanoprinting (a term commonly used to refer to a range of techniques used to transfer a nanoscale pattern to a suitable surface) can also be used for printing NPs and has the potential to lower the cost considerably compared to other nanofabrication methods. However, these techniques are still at an early stage [1, 2].

2.1.1 Top-down processes for the production of NPs

Traditional mechanical grinding methods are able to produce NPs at a relatively low cost from a variety of materials using several milling techniques. This approach is often applied in the production of metallic and ceramic NPs but can also be used to produce complex NPs such as bismuth-telluride-based alloys [3]. Milling involves thermal stress and is energy intensive. Purely mechanical milling can be accompanied by reactive milling in which a chemical or chemo-physical reaction is produced during the milling process. However, the powders obtained show a broad particle size distribution, significant aggregation and impurities from the milling agent.

Physical or chemical exfoliation methods also fall within the top-down category and are often used to obtain laminar nanomaterials such as nanosheets and nano-flakes. In this field, the wide variety of chemical exfoliation methods proposed for the preparation of graphene dispersions should be mentioned. Most are based on the so-called Hummers procedure [4].

2.1.2 Bottom-up processes for the production of NPs

Bottom-up synthesis is used to produce complex nanostructures from atoms to molecules and can often be tailored to obtain good control of the sizes and shapes of NPs. The methods include gas phase and liquid phase reactions. Liquid phase and gas phase synthesis have different time scales; the slower liquid phase processes can be used to obtain thermodynamically controlled products, while for gas phase synthesis kinetic control is often the only option available. Another important distinguishing characteristic between liquid phase and gas phase synthesis is their different behaviour in aggregation: the much higher density and viscosity of the liquid phase slow down NP movement and help to limit aggregation. Moreover, in liquid phase synthesis, NP stability can be further increased through the addition of a stabilizing surface agent.

Gas phase NP production can be divided into methods that operate at atmospheric pressure and those that generate nanocluster (or NP) beams at low pressures [5]. Despite the technical demands and expense of low pressure gas phase methods, they remain a favourite for producing supported NP assemblies. The clusters or NPs can be produced continuously in sizes from 10^5 atoms per cluster down to molecules. Most NP synthesis methods in this category are based on homogeneous nucleation of a supersaturated vapour and subsequent particle growth by condensation and coalescence [6, 7].

Since their initial development, low pressure gas phase methods have differed in how the elemental vapours are generated: arc discharge, pulsed microplasma, laser vaporization, Joule heating and sputtering have been used [8–11]. In general, the formation of the vapour occurs within a reactor at a relatively low pressure. The precursor materials are introduced into the reactor as solids, powders, liquids or gases, where they are evaporated and mixed with a carrier gas. Supersaturated vapour can be produced by cooling, by chemical reaction or by a combination of both. Next, the nucleation process is initiated by the formation of very small cluster embryos from the molecular phase. These nuclei subsequently grow by surface growth mechanisms (heterogeneous condensation, surface reaction) and by collision and coalescence [12–14].

Subsequently, many groups have used these techniques and methods to produce gas phase alloys and core–shell particles with complete flexibility of choice of the core and shell materials [15]. As the goal of this book is not to review the state-of-the-art of inert-gas-condensation sources for NP production, for more details on this topic see [6, 16–19].

In this book, the focus will be entirely on the MS-IGC source because it allows the production of controlled NPs which are suitable for fundamental studies, such as the intriguing field of oxidation processes at the nanoscale. Additionally, these modified sources allow the production of NPs with tailored chemical compositions, fulfilling the increasing need for multifunctional NPs.

2.2 The MS-IGC method

Of the low pressure gas phase techniques, the MS-IGC method has emerged as one of the most flexible and successful, as it allows control of the growth process by varying parameters such as the magnetron power, the length of the so-called *aggregation zone* (figure 2.1) and the inert gas pressure. In addition, the sizes of NPs can be monitored and the composition can be altered *in situ* by the introduction of gases such as oxygen and nitrogen to form oxidized and nitrided NPs with tunable stoichiometries. The wide acceptance of magnetron-sputtering may be a consequence of its ability to generate a large proportion of ionized NPs that can be mass/charge selected by attaching a quadrupole to the source (quadrupole aggregation sources became commercially available in 2001) [20, 21].

The pioneering study published by Haberland and co-workers [11] noted the main advantages of the MS-IGC source as follows. (1) It can produce a very large range of mean NP sizes from 50 to 10^6 atoms per cluster. (2) It has a high degree of ionization, from 30% to 80%, depending on the target materials, allowing easy acceleration of the NPs. (3) A wide variety of elements and alloys can be used as

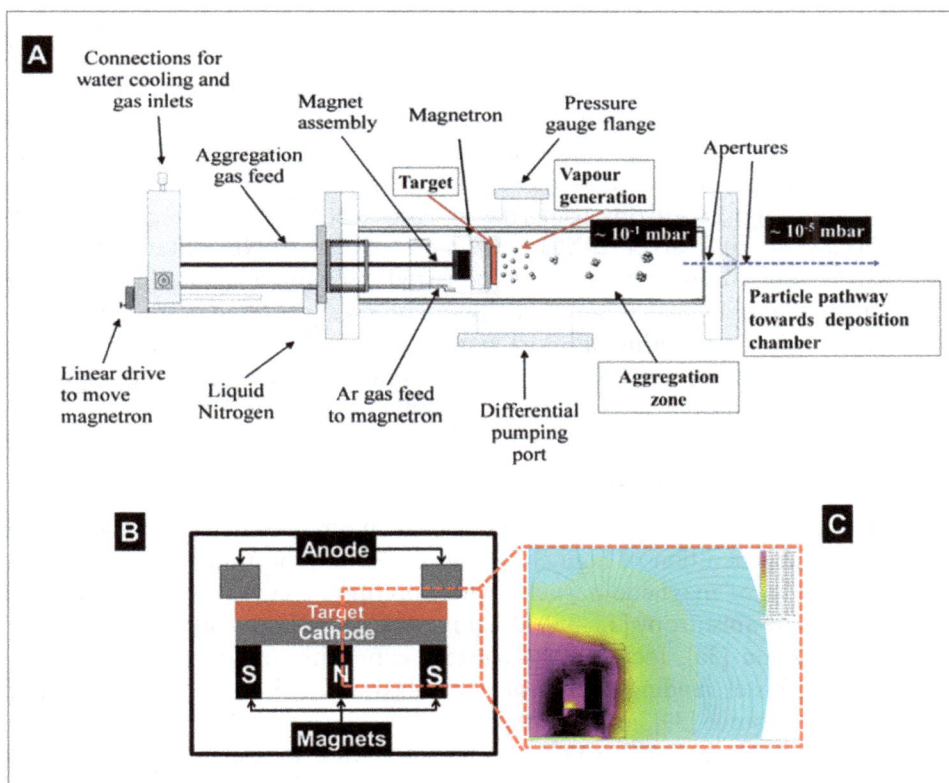

Figure 2.1. (*a*) A diagram of the aggregation zone of a standard MS-IGC source [20]. (*b*) The cathode magnetron geometry. (*c*) The magnetic field configuration in the vertical cross-section of the cathode [21].

source materials. (4) The substrate is not in contact with the plasma and the landing temperature can be regulated.

2.2.1 The synthesis of HNPs

An important research direction in NP synthesis is the expansion from single-component NPs to multicomponent hybrid nanostructures with discrete domains of different materials arranged in a controlled fashion. Progress in nanomaterial synthesis has made possible the mixing of two or more different materials to obtain binary, ternary and multicomponent HNPs. The advantages of multicomponent structures lie in three aspects. The first is to realize multifunctionality; different functionalities can be combined in one single NP. The second advantage is in providing novel functions not available in single-component materials or structures. The third advantage is in achieving enhanced properties due to the couplings between the different components. In fact, because many physical and chemical properties have critical length scales on the order of nanometres, the intimate contact between the nanocomponents in HNPs should allow strong interactions between these components and a possible rational modulation of the physical and chemical properties of each individual component.

In the last decade, further progress in gas phase synthesis was made by sputtering alloyed materials to expand NP composition from single components to hybrid heterostructures [22]. For example, Xu *et al* [23] reported experimental work on iron–silver (Fe–Ag) and cobalt–gold (Co–Au) heterostructured NPs with core–shell and dumbbell-like shapes obtained by the sputtering of FeAg and CoAu composite targets, respectively. Wang reviewed and reported on the preparation of monodispersed FePt NPs with high anisotropy using the gas phase condensation method with an FePt alloyed sputtering target [24]. He *et al* reported $SmCo_5$ NPs obtained using the gas phase condensation method; they prepared a tailored sputtering target by superposing a Co target over a Sm target. Small holes were drilled in the Co target to obtain simultaneous sputtering of both Co and Sm [25]. For other interesting studies of HNPs prepared using the MS-IGC method from single alloyed sputtering targets such as Co–Au, Mo–Cu and Au–Pd see [26–29].

2.2.2 Modification of the MS-IGC source

Several modifications of gas phase sources have been reported in the literature. Most are aimed at the generation of high cluster fluxes [30–32], post-annealing treatments for the generation of magnetic particles with appropriate crystallographic structure and high magnetic anisotropy [33–35]. Other modifications have addressed the control of cluster oxidation [36], focusing the cluster beam with aerodynamic lenses [37], control of the cluster landing energy on the substrates [38] and room temperature operation of the source [39], among others.

Nevertheless, only a few works have been reported in the literature on modifications of the NP source to allow the *in situ* control and adjustment of the chemical composition or the position of each element within the generated HNP. For example, Pellarin *et al* used a double ablation laser to generate $(C_{60})_n Si_m$ cationic clusters by quenching the vapours from two independent C_{60} and Si targets [40, 41], while Sumiyama *et al* used independent sputtering sources to produce Co–Si and Fe–Si core–shell clusters [42]. In both designs, the positions of the targets were fixed and hence could not be moved to adjust the length of the aggregation zone, as is customary in more standard NP sources [31, 43]. Otherwise, in most cases the chemical compositions of the NPs produced depended on the chemical composition of the single target material. Modification of the chemical composition of the NPs required the substitution of the single target by another with a different chemical composition.

Recently, Martinez *et al* [44] reported on the use of two or more independent target materials to produce AgAuPd alloyed particles using the MS-IGC method. In fact, they modified their system to generate NPs with a controllable and tunable chemical composition while maintaining control of the cluster size. In their patented design [45], they replaced the typical two-inch magnetron with three one-inch independent magnetrons whose powers and positions could be controlled individually.

Alternatively, the Mantis Deposition company [21] has started commercializing the patented Nanogen Trio magnetron sputtering cathode [46], the details of which are presented in figure 2.2. A photograph of the Nonogen Trio showing the three

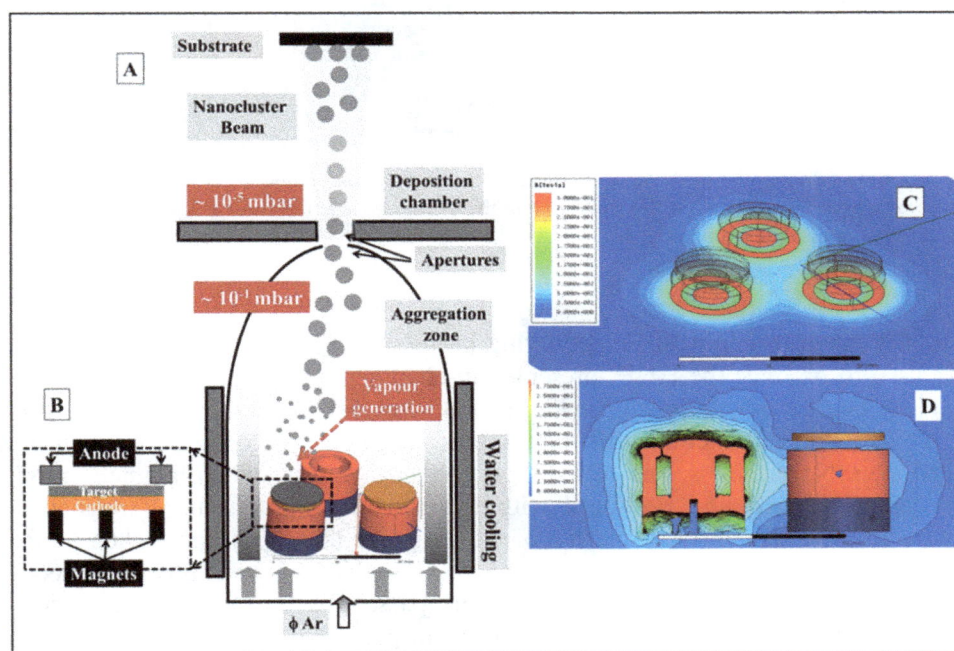

Figure 2.2. (*a*) A diagram of the modified NP source used to produce the HNPs presented in this work. (*b*) shows the three independent magnetron-sputtering cathodes; each target is one inch in diameter. The distribution of the magnetic field in the half-height plane crossing the three cathodes of the sputtering trio is plotted in (*c*). The magnetic field configuration in the vertical cross-section of one cathode is shown in (*d*). The magnetic field was calculated using the finite-element software package Maxwell 3d v14.

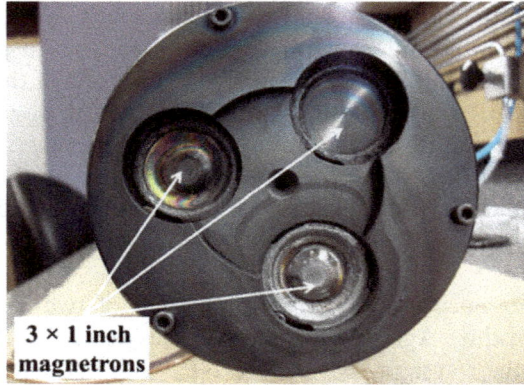

Figure 2.3. A photograph showing the three independent magnetron cathodes integrated in a single sputtering head as shown in figure 2.2 [46].

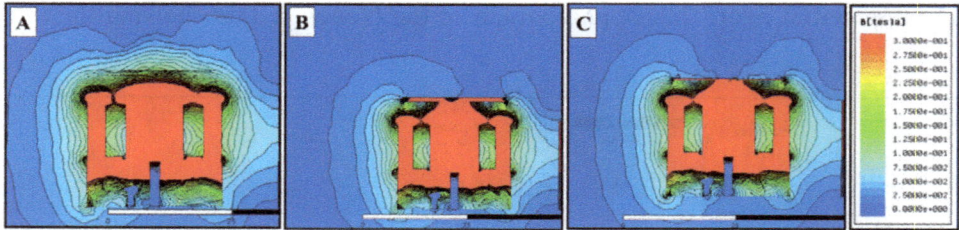

Figure 2.4. The distribution of the magnetic field in the vertical cross-section of the magnetron sputtering cathodes (*a*) without an Fe target, (*b*) with a 1 mm thick Fe target and (*c*) with a 0.3 mm thick Fe target. The target is one inch in diameter.

cathodes integrated in a single sputtering head is shown in figure 2.3. This system was used to produce the HNPs presented in chapters 3 and 4 of this book.

These latest modifications opened a new horizon for the design and preparation of HNPs with precise control of their composition, morphology and size, all in a single step. Nevertheless, one of the most challenging tasks when magnetic HNPs are being designed and produced with this novel configuration of magnetrons (figure 2.3) is the efficient sputtering of the magnetic targets. In fact, when a magnetic target is placed on the cathode, it provides a path of magnetic flux generated by the permanent magnet of the cathode. Flux passes through it and only a small portion escapes and reaches the surface of the target to contribute to the generation of magnetron plasma. For this reason, as shown in figure 2.4, a study of the effects of the thickness of the iron target on the magnetic field distribution near the surfaces of the individual magnetrons was performed using a finite-element software package (Maxwell 3d v14). For adequate sputtering, the thickness of the Fe target should not exceed 1 mm.

2.3 Factors influencing the formation of HNPs using gas phase methods

For both fundamental and technological purposes, new developments are strongly dependent on accurate control of the growth of NPs. It is essential to control the chemical composition, crystallinity, morphology, size and size distribution to tailor the physical and chemical properties of the NPs. Success in obtaining tailored HNPs by a physical route encompasses various factors such as the miscibility and surface energy of the materials and the interfacial energy between the substrate and the NPs.

In fact, in the case of bimetallic NPs, for example, experimental studies and theoretical modelling have determined that the segregation, mixing and chemical ordering of metals within a bimetallic system is a result of an interplay between different intrinsic and extrinsic properties of the metals involved as follows [47–49]:

- *Bond strength*. For two different metals A and B, if the A–B bond is stronger than the A–A and B–B bonds, this will favour the inter-mixing of elements.
- *NPs surface energies*. The material with the highest surface energy tends to occupy core positions.
- *Atomic radii*. For immiscible materials, the element with the smallest atomic radius tends to occupy the core position. However, for example, if two metals A and B are fully or partially miscible, the formation of either a solid solution or a new phase can occur. This depends on the position of the elements in the periodic table, their relative concentrations and the temperature of the alloy. In fact, for a substitutional solid solution (atoms of A replace atoms of B or vice versa, depending on the relative concentrations of A and B) a necessary (but not sufficient) condition to accommodate an appreciable number of atoms of A in the B lattice is that the difference in atomic radii between the two elements is less than 15%. Otherwise, a new phase will form. In contrast, for an interstitial type of solid solution, the atomic radius of atom A must be substantially smaller than that of atom B.
- *Electronegativity and charge transfer*. When the elements involved in a reaction have a large electronegative difference, the probability that they will form an intermetallic compound instead of a solid solution is higher.

Other factors that can affect the formation of bimetallic NPs are the crystal structure and the valence of each element. Copper and nickel exemplify a substitutional solid solution. These two elements are completely miscible at all proportions. With regard to the above rules that govern the degree of miscibility, the atomic radii for copper and nickel are 0.128 and 0.125 nm, both have a fcc crystal structure and their electronegativities are 1.9 and 1.8, respectively. In addition, the most common valences are +1 for copper (although it can sometimes be +2) and +2 for nickel. An interstitial solid solution is formed by carbon when added to iron; the maximum concentration of carbon is approximately 2%. The radius of the carbon atom is much smaller than that of iron: 0.071 nm versus 0.124 nm [49].

Regarding examples of non-metallic HNPs prepared using the MS-IGC method, two examples of binary NPs are presented in chapter 3. The first addresses intermetallic FeAl NPs where the components are bulk-miscible (section 3.1). The second example discusses the metallic–dielectric Ag–Si binary NPs in which the materials are bulk-immiscible (section 3.2). The case of the ternary NPs Fe–Ag–Si is presented in chapter 4.

References

[1] Zheng Y, Lalander C H, Thai T, Dhuey S, Cabrini S and Bach U 2011 Gutenberg-style printing of self-assembled nanoparticle arrays: electrostatic nanoparticle immobilization and DNA-mediated transfer *Angew. Chem. Int. Edn Engl.* **50** 4398–402

[2] Engstrom D S, Porter B, Pacios M and Bhaskaran H 2014 Additive nanomanufacturing—a review *J. Mater. Res.* **29** 1792–816

[3] Takashiri M, Tanaka S, Takiishi M, Kihara M, Miyazaki K and Tsukamoto H 2008 Preparation and characterization of $Bi_{0.4}Te_{3.0}Sb_{1.6}$ nanoparticles and their thin films *J. Alloys Compounds* **462** 351–55

[4] Hummers W S and Offema R E 1958 Preparation of graphitic oxide *J. Am. Chem. Soc.* **80** 1339

[5] Binns C 2009 *Handbook of Metal Physics: Metallic Nanoparticles* ed J Blackman (Amsterdam: Elsevier) ch 3

[6] Swihart M T 2003 Vapor-phase synthesis of nanoparticles *Curr. Opin. Colloid Interface Sci.* **8** 127–33

[7] Kruis F E, Fissan H and Peled A 1998 Synthesis of nanoparticles in the gas phase for electronic, optical and magnetic applications—a review *J. Aerosol Sci.* **29** 511–35

[8] Milani P and deHeer W A 1990 Improved pulsed laser vaporization for production of intense beams of neutral and ionized clusters *Rev. Sci. Instrum.* **61** 1835–38

[9] Siekmann H R, Lüder Ch, Faehrmann J, Lutz H O and Meiwes-Broer K H 1991 The pulsed arc cluster ion source *Z. Phys. D* **20** 417–20

[10] Haberland H, Karrais M and Mall M 1991 A new type of cluster and cluster ion source *Z. Phys. D* **20** 413–15

[11] Haberland H, Karrais M, Mall M and Thurner Y 1992 Thin films from energetic cluster impact: A feasibility study *J. Vac. Sci. Technol. A* **10** 3266–71

[12] Granqvist C G and Buhrman R A 1976 Ultrafine metal particles *J. Appl. Phys.* **47** 2200–19

[13] Sattler K, Mühlbach J and Recknagel E 1980 Generation of Metal Clusters Containing from 2 to 500 Atoms *Phys. Rev. Lett.* **45** 821–24

[14] Muhlbach J, Pfau P, Recknagel E and Sattler K 1981 Cluster emission from the surfaces of Bi, Sb and Se *Surf. Sci.* **106** 18–26

[15] Pauly H 2000 *Atom, Molecule and Cluster Beams II (Springer Series on Atomic, Optical, and Plasma Physics* vol 32) (Berlin: Springer)

[16] Xirouchaki C and Palmer R E 2004 Deposition of size-selected metal clusters generated by magnetron sputtering and gas condensation: a progress review *Philos. Trans. R. Soc. A* **362** 117–24

[17] Bansmann J *et al* 2005 Magnetic and structural properties of isolated and assembled clusters *Surf. Sci. Rep.* **56** 189–275

[18] Wegner K, Piseri P, Vahedi-Tafreshi H and Milani P 2006 Cluster beam deposition: a tool for nanoscale science and technology *J. Phys. D: Appl. Phys.* **39** R439–59

[19] Binns C *et al* 2005 The behaviour of nanostructured magnetic materials produced by depositing gas-phase nanoparticles *J. Phys. D: Appl. Phys.* **38** R357–79

[20] Oxford Applied Research www.oaresearch.co.uk/oaresearch/cluster/

[21] Mantis Deposition www.mantisdeposition.com, private communication

[22] Yin F, Wang Z W and Palmer R E 2011 Controlled formation of mass-selected Cu–Au core–shell cluster beams *J. Am. Chem. Soc.* **133** 10325–27

[23] Xu Y H and Wang J P 2008 Direct Gas-Phase Synthesis of heterostructured nanoparticles through phase separation and surface segregation *Adv. Mater.* **20** 994–9

[24] Wang J P 2008 Magnetic nanoparticles and their assembly for future magnetic media *Proc. IEEE* **96** 1847–63

[25] He S, Jing Y and Wang J P 2013 Direct synthesis of large size ferromagnetic $SmCo_5$ nanoparticles by a gas-phase condensation method *J. Appl. Phys.* **113** 134310

[26] Krishnan G, Verheijen M A, Ten-Brink G H, Palasantas G and Kooi B J 2013 Tuning structural motifs and alloying of bulk immiscible Mo–Cu bimetallic nanoparticles by gas-phase synthesis *Nanoscale* **5** 5375–83

[27] Perez-Tijerina E, Pinilla M G, Mejia-Rosales S, Ortiz-Mendez U, Torres A and Jose-Yacaman M 2008 Highly size controlled synthesis of Au/Pd nanoparticles by inert-gas condensation *Faraday Discuss.* **138** 353–62

[28] Llamosa Perez D, Espinosa A, Martinez L, Roman E, Ballesteros C, Mayoral A, Garcia-Hernandez M and Huttel Y 2013 Thermal diffusion at nanoscale: from CoAu alloy nanoparticles to Co@Au core/shell structures *J. Phys. Chem. C.* **117** 3101–8

[29] Mayoral A, Mejia-Rosales S, Mariscal M, Perez-Tijerina E and Jose-Yacaman M 2010 The Co–Au interface in bimetallic nanoparticles: a high resolution STEM study *Nanoscale* **2** 2647–51

[30] Imanaka M, Katayama T, Ohshiro Y, Watanabe S, Arai H and Nakagawa T 2004 Nanocluster ion source by plasma-gas aggregation *Rev. Sci. Instrum.* **75** 1907–9

[31] Iles G N, Baker S H, Thornton S C and Binns C 2009 Enhanced capability in a gas aggregation source for magnetic nanoparticles *J. Appl. Phys.* **105** 024306

[32] Momin T and Bhowmick A 2010 A new magnetron based gas aggregation source of metal nanoclusters coupled to a double time-of-flight mass spectrometer system *Rev. Sci. Instrum.* **81** 075110

[33] Stoyanov S, Huang Y, Zhang Y, Skumryev V, Hadjipanayis G C and Weller D 2003 Fabrication of ordered FePt nanoparticles with a cluster gun *J. Appl. Phys.* **93** 7190

[34] Qiu J-M, Xu Y H, Judy J H and Wang J-P 2005 Nanocluster deposition for high density magnetic recording tape media *J. Appl. Phys.* **97** 10P704

[35] Bai J and Wang J P 2005 High-magnetic-moment core–shell-type FeCo–Au/Ag nanoparticles *Appl. Phys. Lett.* **87** 152502

[36] Kennedy M K, Kruis F E, Fissan H, Mehta B R, Stappert S and Dumpich G 2003 Tailored nanoparticle films from monosized tin oxide nanocrystals: Particle synthesis, film formation, and size-dependent gas-sensing properties *J. Appl. Phys.* **93** 551–60

[37] Fonzo F D *et al* 2000 Focused nanoparticle-beam deposition of patterned microstructures *Appl. Phys. Lett.* **77** 910

[38] Goldby I M, Issendorff B V, Kuipers L and Palmer R E 1997 Gas condensation source for production and deposition of size-selected metal clusters *Rev. Sci. Instrum.* **68** 3327

[39] Majumdar A, Köpp D, Ganeva M, Datta D, Bhattacharyya S and Hippler R 2009 Development of metal nanocluster ion source based on dc magnetron plasma sputtering at room temperature *Rev. Sci. Instrum.* **80** 095103

[40] Pellarin M, Ray C, Lermé J, Vialle J L, Broyer M and Mélinon P 2000 Gas phase study of silicon-C_{60} complexes: Surface coating and polymerization *J. Chem. Phys.* **112** 8436–45

[41] Masenelli B, Tournus F, Mélinon P, Blasé X, Perez A, Pellarin M and Broyer M 2004 Nanostructured films from $(C_{60})_n Si_m$ clusters *Appl. Surf. Sci.* **226** 226–30

[42] Sumiyama K, Hihara T, Liang P D and Katoh R 2005 Structure and magnetic properties of Co/CoO and Co/Si core–shell cluster assemblies prepared via gas-phase *Sci. Technol. Adv. Mater.* **6** 18–26
Sumiyama K, Katoh R, Kadowaki S and Hihara T 2010 Fe–Si core/Si-shell clusters prepared by double glow discharge sources *J. Nanopart. Res.* **12** 2589–96

[43] Quesnel E, Pauliac-Vaujour E and Muffato V 2010 Modeling metallic nanoparticle synthesis in a magnetron-based nanocluster source by gas condensation of a sputtered vapour *J. Appl. Phys.* **107** 054309

[44] Martínez L, Díaz M, Román E, Ruano M, Llamosa D and Huttel Y 2012 Generation of Nanoparticles with Adjustable Size and Controlled Stoichiometry: Recent Advances *Langmuir* **28** 11241–9

[45] Román-García E L, Martínez-Orellana L, Díaz-Lagos M and Huttel Y 2010 Dispositivo y procedimiento de fabricación de nanoparticulas *Spanish Patent specification* P201030059, PCT/ES2011/070032

[46] Allers L 2013 Production of nanoparticles *US Patent Application* 20130270106

[47] Paz-Borbon L O 2011 *Computational Studies of Transition Metal Nanoalloys (Springer Theses)* (Berlin: Springer)

[48] Langlois C, Alloyeau D, Bouar Y L, Loiseau A, Oikawa T, Mottet C and Ricolleau C 2008 Growth and structural properties of CuAg and CoPt bimetallic nanoparticles *Faraday Discuss.* **138** 375–91

[49] Callister W D Jr 2007 *Materials Science and Engineering: An Introduction* 7th edn (Hoboken, NJ: Wiley)

Chapter 3

Designing binary nanoparticles

3.1 An introduction to binary NPs

The synthesis and assembly of NPs of two different materials into a binary NP can provide a general and inexpensive path to a large variety of materials (metamaterials) with precisely controlled chemical composition and tight placement of components. As discussed in section 1.2, binary NPs can be combined in different configurations: 1) random or ordered alloys in a mixed state, 2) core–shell morphology (including multishell structures), 3) multicore–shell morphology and 4) dumbbell-like and satellite morphologies in segregated or phase-separated states [1–3].

Despite the advantages that the MS-IGC method offers in producing phase-separated NPs with free-residual contamination both at the interface between the different components and on the surface, only a few studies have reported the formation of different morphological and structural configurations for the same type of NP with a fixed chemical composition [1, 4].

Success in obtaining binary NPs with controlled structural and morphological configurations by the physical route depends strongly on the miscibilies and surface energies of the elements involved in the reaction [5]. The factors affecting NP formation are discussed in more detail in section 2.3. In this chapter, two examples of binary NPs are presented. The first example addresses the synthesis and characterization of intermetallic FeAl NPs whose components are partially bulk-miscible. The resulting NPs consist of a nanoalloy of an iron aluminide crystalline core (DO3 phase) encapsulated in an alumina shell that reduces inter-particle interactions and also prevents further oxidation and segregation of the FeAl core. The second example addresses the synthesis and characterization of metallic–dielectric Ag–Si binary NPs, whose materials are bulk-immiscible. In this case, appropriate tuning of the sputtering power ratio of the Si and Ag independent targets allows the production of NPs with distinctive multicore–shell and core–satellite morphologies. The NPs obtained in both cases demonstrate that the modification of the MS-IGC system makes it possible to overcome the limitations that other deposition techniques face regarding nanoalloy oxidation and the immiscibility of materials.

3.2 The synthesis and characterization of FeAl HNPs

3.2.1 Introduction

Current advances in nanotechnology rely increasingly on multifunctional nano-structures. Within this class of materials, soft magnetic alloys at the nanoscale level have generated intense interest because of their possible uses in many technological applications such as power transformers, inductive devices and magnetic sensors [6, 7].

For this reason, aluminides of transition metals (e.g. nickel and iron) are promising materials due to their excellent oxidation resistance. Specifically, the FeAl system easily and quickly forms an adherent oxide layer on its surface that provides high protection to the underlying alloy when exposed to oxidizing and corrosive environments. This property, along with soft magnetic behaviour, excellent thermal conductivity and low mass density, make this material a very promising candidate for many bio- and nanotechnology applications. However, traditional synthesis methods must overcome the long-standing challenge of developing precise nanoalloys with effective morphological control and stability against oxidation. In fact, when bimetallic systems are considered at the nanoscale, oxidation, phase segregation and aggregation due to inter-particle magnetic interactions are expected, resulting in the alteration of magnetic properties and raising the question of the feasibility of soft magnetic nanoalloys [8].

The fabrication of iron aluminide alloy NPs using the MS-IGC method is described below. Note the advantage of working in a vacuum, which allows a better control of the chemical purity of the NPs, specifically the oxidation process of NPs. The desired chemical composition of the NPs can be obtained by tuning the magnetron power applied on each target (Fe and Al) independently while they are co-sputtered. When the NPs are exposed to air, oxidation occurs at high rates, resulting in a pure external shell of alumina.

3.2.2 Deposition of FeAl NPs

FeAl NPs were produced using the modified MS-IGC system as described in section 2.2. A supersaturated vapour of metal atoms was generated by co-sputtering iron and aluminium from two independent neighbouring targets in an argon (Ar) atmosphere. The aggregation chamber was water-cooled and evacuated to $\sim 10^{-8}$ mbar prior to sputtering. Fe (99.9%) and Al (99.9995%) targets one inch in diameter were used in the dc co-sputtering process. The working pressure was maintained at 3×10^{-1} mbar in the aggregation zone. The pressure in the deposition chamber was set at 8.4×10^{-4} mbar. This differential pressure is a key factor; it determines the residence time of the NPs in the aggregation zone and therefore the crystallinity, size and shape of the NPs (the effects of residence time on the formation of NPs are discussed in section 4.5). The dc powers applied to the Fe and Al targets were 11 W and 16 W, respectively. The power ratio was fixed to work in the Fe-rich part of the Fe–Al binary phase diagram in which the DO3 and A2 phases are growing and stable, respectively, at temperatures less than 500 °C. The aggregation zone length was set to 90 cm. The NPs were deposited on silicon substrates and silicon nitride TEM grids for characterization. At the end of the deposition process, the NPs were removed from the deposition chamber and exposed to air for a few minutes before the characterization process was initiated.

3.2.3 Characterization methods

The size, morphology and crystal structure of the intermetallic NPs were examined using an image-corrected scanning/transmission electron microscope FEI Titan 80–300 kV operated at 300 kV (see figure 1.8). The chemical composition and surface oxidation of the NPs were evaluated using the XPS system Kratos Axis UltraDLD 39-306 equipped with a mono-AlKα source operated at 300 W (see figure 1.9).

3.2.4 The morphology, structure and composition of FeAl NPs

Low magnification and high magnification TEM images (figures 3.1(a) and (b), respectively) show the core–shell morphology of the NPs obtained. Figure 3.1(f) shows the size probability distribution function (PDF) calculated over 100 NPs from the TEM images. This figure reveals that approximately 42% of the NPs were 8–10 nm in size and 30% were 10–12 nm in size. The remaining 30% were 12–16 nm in size. At this point it is important to note that no size selection of NPs was applied in this study. The deposition system used in this study is equipped with a quadrupole mass filter placed between the aggregation zone and the deposition chamber; however, for this study, it was turned off. The HRTEM image of a representative NP (figure 3.1(c)) shows its distinctive core–shell morphology with a spherical configuration. From the HRTEM images, the shell thickness was determined to be in the range of 2–3 nm. The inter-planar distance estimated from the lattice fringes was found to be 2.03 Å, which can be assigned to the Fe-rich A2, the B2, or the DO3 phase. However, the high temperature ordered B2 phase was not expected in this case due to the relatively low temperatures in the inert-gas-condensation technique [10]. Furthermore, the fast Fourier transform (FFT) diffractogram (figure 3.1(d)) corresponding to the HRTEM image in figure 3.1(c), combined with the simulated electron diffraction pattern oriented in the [00-1] zone axis (figure 3.1(e)) calculated using ^{TM}Crystal Maker software, clearly demonstrates the DO3 nature of the NP's core crystal structure.

To obtain further information on the chemical composition of these NPs, EELS analysis was performed (figures 3.2(c) and (d)). The EELS spectra obtained from areas 1, 2 and 3 as indicated in figure 3.2(b) show the presence of Fe, Al and O within the NP. The spectra corresponding to area 2 with brighter contrast within the NP (figure 3.2(b)) show the edges of both the Fe–$L_{2,3}$ at 707 eV (figure 3.2(c)) and Al–$L_{2,3}$ at 76 eV (figure 3.2(d)). The EELS spectra corresponding to the shell (area 1 and area 3) are Al- and O-rich (O–K 532 eV) areas, indicating that the shell is dominated by Al and O components.

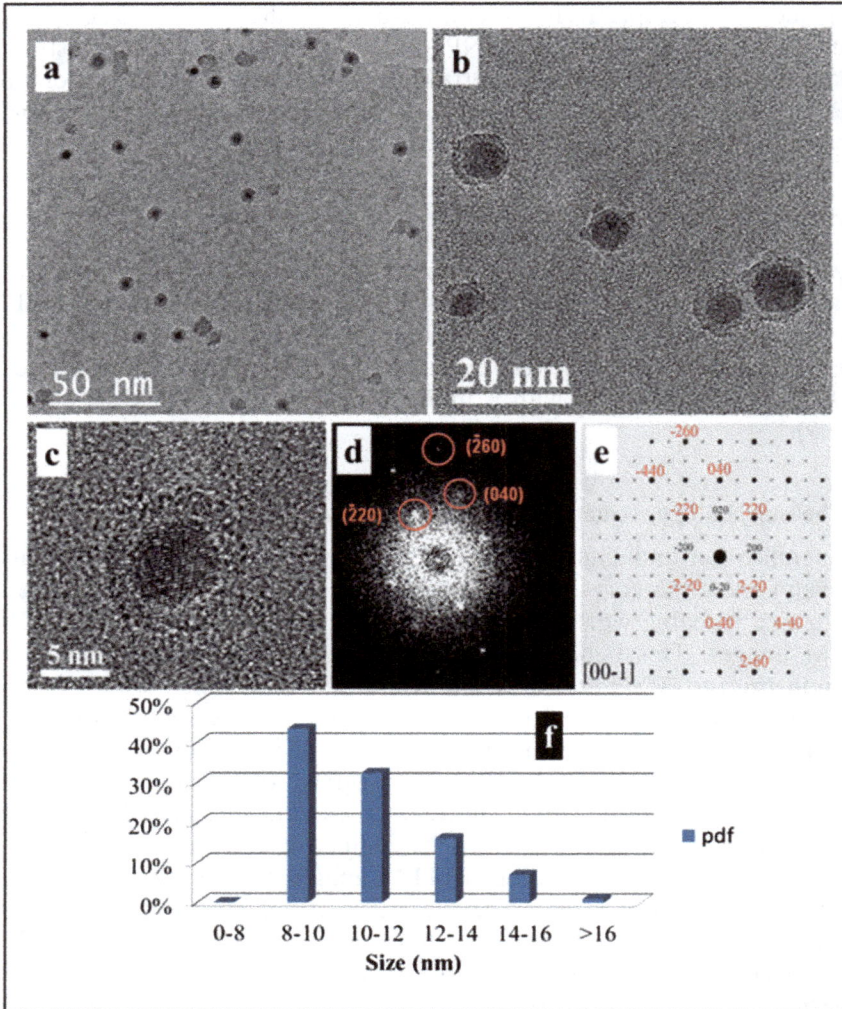

Figure 3.1. (*a*), (*b*) Low and high magnification TEM images showing the core–shell morphology of the NPs. (*c*) A HRTEM image showing the single crystalline core encapsulated with an amorphous shell, (*d*) the corresponding FFT and (*e*) the electron diffraction pattern in the [00-1] zone axis orientation calculated using ᵀᴹCrystal Maker software. (*f*) The size PDF of the NPs. (*c*), (*d*) and (*e*) are reproduced from [2] under CC BY 3.0.

3.2.5 An XPS study of FeAl HNPs

XPS core level spectra of Al 2p and Fe 2p are shown in figures 3.3(*a*) and (*b*), respectively. The results show that Fe and Al are present both in the metallic (73.5 eV and 706.8 eV) and the oxide (74.4 eV and 710.4 eV) states, respectively. The ratio between the peak areas of metallic Al 2p (73.5 eV) and Fe 2p (706.8 eV) is ~27%, corresponding to the DO3 phase ($Fe_{73}Al_{27}$) in the binary phase diagram of

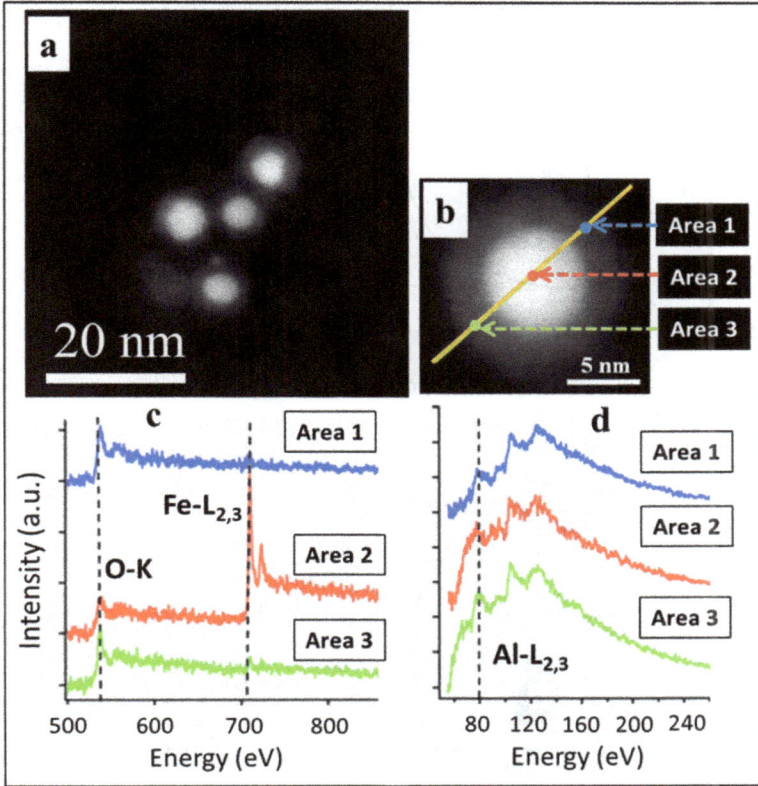

Figure 3.2. (*a*), (*b*) Low and high magnification STEM images showing the core–shell morphology of the NPs. (*c*), (*d*) EELS spectra obtained from different areas of a representative NP (as indicated in (*b*)). The EELS spectra of areas 1, 2 and 3 show the presence of Fe, Al and O within the NP. As seen in (*c*), area 2 shows a strong intensity of the Fe–$L_{2,3}$ edge corresponding to the position of the bright core, whilst the spectra on either side of the shell (areas 1 and 3) are dominated by the Al–$L_{2,3}$ and O–K edges. (*b*), (*c*) and (*d*) are adapted from [2] under CC BY 3.0.

iron aluminide. Moreover, the peak corresponding to metallic Al (figure 3.3(*a*)) is shifted towards higher binding energy (73.4 eV instead of 72 eV), suggesting that Al atoms coordinate with Fe atoms. This matches the reported value for the Fe_3Al phase [11, 12] exactly. The peak at 75.3 eV binding energy (figure 3.3(*a*)) is an indication of the formation of Al_2O_3 on the surface.

The same conclusion can be drawn from the O 1s peak (figure 3.4(*a*)) at 532.97 eV, which corresponds to the reported value for Al_2O_3 [12].

In addition, the deconvolution of the Fe 3p peak into Fe^{2+} and Fe^{3+} peaks with an atomic ratio of 1 : 2 (figure 3.4(*b*)), in combination with the Al 2p peak at 74.4 eV (figure 3.3(*a*)) and the O 1s peak at 531.57 eV (figure 3.4(*a*)), suggests the presence of spinel oxide $FeAl_2O_4$ in the inert shell after exposure to air [13, 14]. These results are in full agreement with the study reported by Graupner and co-workers [11]. In fact, when extracted from the deposition chamber and exposed to air, the NPs are

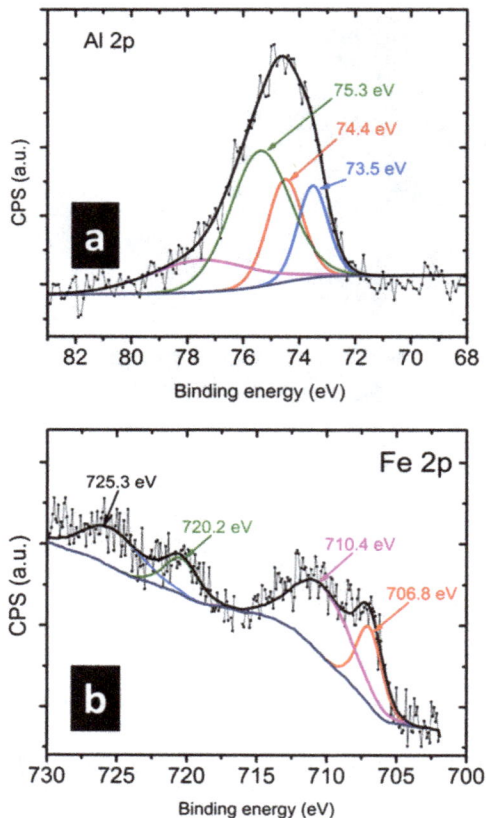

Figure 3.3. XPS spectra corresponding to (*a*) Al 2p and (*b*) Fe 2p core levels from the NPs. For the measurements, a high coverage sample of FeAl NPs deposited on Si was prepared and transferred to the XPS analysis chamber after a few minutes of exposure to the ambient atmosphere.

oxidized at high rates and all Al atoms available in the near-surface region will react with oxygen from the air to form Al_2O_3. At ambient temperatures, however, the diffusion rate of aluminium from the core to the surface of the NPs is too low to compensate the loss of Al at the surface. The same might hold for any other mechanism involved in Al transport to the surface. Therefore, Fe-enriched regions will be exposed to the incoming oxygen and, in the absence of competition from neighbouring Al atoms, the Fe will be oxidized as well. This effect should be more pronounced when the transport of Al to the surface is slower.

The magnetic behaviour of these NPs and their stability against oxidation are presented in the original work of [2]. Briefly, the NPs show soft magnetic properties with a high saturation magnetization (170 emu g^{-1}) and low coercivity (less than 20 Oe) at room temperature. Moreover, FeAl NPs demonstrate good stability against further oxidation after 30 days of exposure to air.

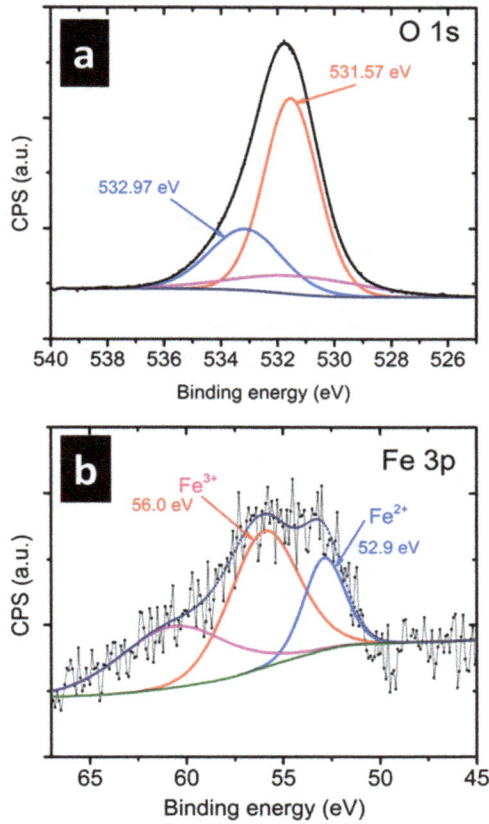

Figure 3.4. XPS spectra and curve fittings for the (*a*) O 1s region and (*b*) Fe 3p region of the sample in figure 3.3.

3.2.6 The formation mechanism

The size, morphology and crystal phase of the NPs can be modulated by controlling the experimental conditions during deposition. In particular, adjusting the power of each target within the co-sputtering process allows control of the plasma densities close to the target surface; this plays an important role in the nucleation, growth and crystallization of the NPs. In the particular experimental conditions described in this section, the low plasma density near the surface of the Fe target results in just partial nucleation and growth afterwards, when the atoms fly into the aggregation zone (see figure 2.2), leading to the formation of small, amorphous Fe nanoclusters. For further details of this process, see section 4.7 [1, 15, 16]. For Al atoms, the high concentration of electrons and ions due to the high sputtering yield induce high-density plasma near the surface of the Al target, which leads to the complete nucleation and growth of the NPs before they leave the plasma region. Subsequently, Fe and Al nanoclusters collide and coalesce with each other in the aggregation zone, forming large NPs (figure 3.5, step 2). According to the studies

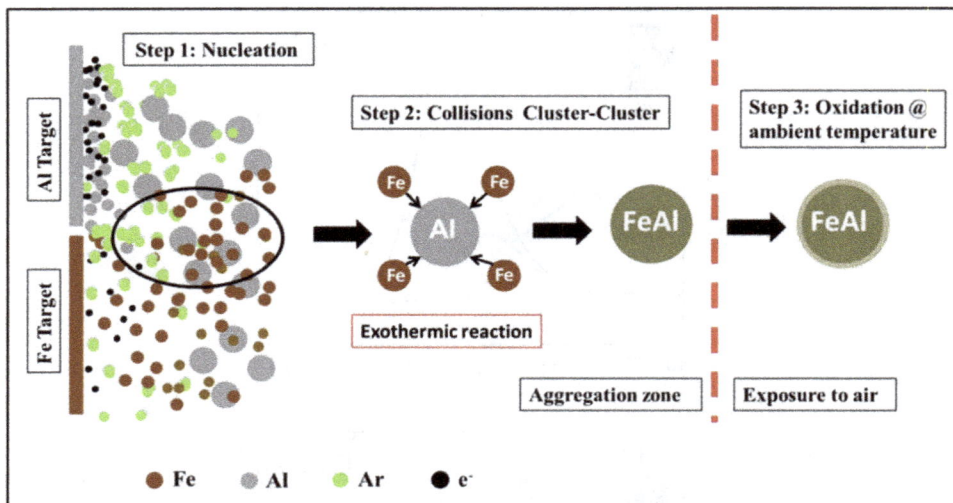

Figure 3.5. A diagram summarizing the suggested mechanism of formation of the FeAl@Al$_2$O$_3$ NPs. Step 1 shows the nucleation and growth phase in the plasma zone near the surface of the Fe and Al sputtered targets. Step 2 represents cluster–cluster collisions: the Fe and Al nanoclusters are mixed in an exothermic reaction and a crystalline FeAl alloy is formed. After exposure to air, high rate oxidation is produced, resulting in an Al$_2$O$_3$ protective shell.

reported by Yang and co-workers [17, 18], a core–shell structure is expected, with segregation of Al at the surface in most cases. However, in this case, upon exiting the plasma zone there are two competing mechanisms, the rates of which dictate the morphology of the resultant NPs. On the one hand, the important negative free enthalpy of mixing of Fe and Al [19] induces an exothermic reaction which, combined with subsequent inter-cluster collisions, increases and maintains the high temperature of the Al nanoclusters for a period of time sufficient to create the Fe–Al crystalline alloy. On the other hand, collisions with inert gas (argon) atoms in the aggregation zone tend to cool the nanoclusters faster and consequently stabilize the single crystalline structure. When the former process is practically complete, the latter takes over. Finally, when exposed to air, the NPs are oxidized at high rates and all Al atoms available in the near-surface region will react with oxygen from the air to form Al$_2$O$_3$. Oxidation at a low rate in the main chamber might also occur, leading to the formation of an alumina shell; however, the formation of iron oxide (XPS spectra Fe 2p and Fe 3p) indicates a high oxidation rate mechanism [11].

3.3 The synthesis and characterization of AgSi NPs

3.3.1 An overview of AgSi NPs

AgSi NPs are of significant interest because combined metal–dielectric (or metal–semiconductor) systems exhibit both photoluminescence and plasmonic properties, which make them attractive candidates for applications ranging from biosensors to informatics and storage devices. Regarding plasmonic properties, interesting studies have been reported on the tuning of the so-called local-surface-plasmon-resonance peak position based on Ag inter-particle coupling and the high refractive index of the dielectric matrix [20–22]. In addition, in the field of bioapplications, it has been demonstrated that the use of pure silver NPs gives rise to inconvenient liberation of Ag ions, which are toxic for living cells [23]. However, as reported by Sotiriou and co-workers [24], for example, after coating the Ag NPs with a continuous silica shell approximately 2 nm thick, the release of Ag toxic ions decreases from 12 mg L^{-1} for uncoated silver NPs to less than 2 mg L^{-1} for silica-coated NPs; the percentage of dead cells then declines from ~70% for uncoated NPs to ~10% for coated NPs.

Excluding a few exceptions, the most commonly reported techniques for synthesizing AgSi NPs are limited to wet chemical methods. For example, Aslan and co-workers reported the controlled growth of a silica shell on Ag colloids for bio-imaging applications. Their process consisted of two steps: first, silver colloids were prepared by reduction of silver nitrate by sodium citrate and, second, silica shells of various thicknesses were grown on the Ag colloids [25]. Chen and co-workers reported a modified Stober method for the synthesis of SiO_2–Ag core–satellite NPs, which can be used for plasmonic-based chemical sensing and photocatalysis applications [26]. In contrast, only a few attempts to synthesize AgSi NPs using physical methods have been reported. For example, Nomoev and Bardakhanov reported Ag–Si core–shell NPs prepared using the electron-beam evaporation method [23]. Mertens, in his PhD thesis, reported the synthesis of 2D NP arrays embedded in an amorphous silicon (a-Si) matrix by sequential Si/Ag/Si electron-beam evaporation of Si and Ag [27].

The immiscibility of Si and Ag in the bulk solid state illustrates the limitations of single alloy targets which limit AgSi NP production from a single alloyed target [28, 29]. Here again, this limitation can be overcome with the modified MS-IGC system described in figure 2.2. The experimental details of the synthesis and characterization of the AgSi NPs are presented in sections 3.3.2 and 3.3.3, respectively. In section 3.3.4, a formation mechanism for these NPs is suggested, based on the combination of the experimental conditions and the physico-chemical properties of the materials (mainly the miscibility and surface energy).

3.3.2 Deposition and characterization methods for AgSi HNPs

The AgSi NPs were synthesized by co-sputtering Si and Ag from two independent targets located side-by-side on an integrated magnetron sputtering head, as shown in figures 2.2 and 2.3. The volume fraction of each component was controlled by tuning the magnetron power applied to each target independently. For all depositions, the Ar flow rate was set to 80 sccm, resulting in an aggregation zone pressure reading of 3×10^{-1} mbar. The aggregation zone length was set to 90 mm. The base pressure (7.0×10^{-8} mbar) in the main deposition chamber was increased to 3.5×10^{-3} mbar by introducing Ar flux (50 sccm). By reducing the pressure differential between the aggregation zone and the deposition chamber, the residence time of the NPs in the aggregation zone was increased, leading to the formation of larger NPs with a higher number of cores and thicker Si shells [15, 30]. The Si target sputtering power was fixed at 60 W, and the tuning of the particle composition and morphology was

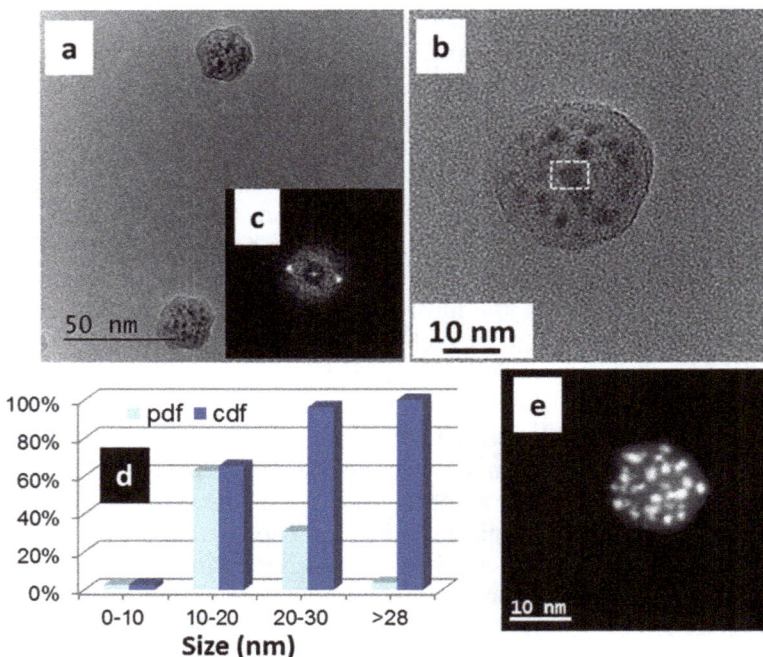

Figure 3.6. (*a*) A high magnification TEM image showing the multicore–shell morphology of the AgSi NPs. (*b*) A HRTEM image of a representative AgSi multicore–shell NP and (*c*) the FFT patterns corresponding to an individual core as indicated in (*b*). The d-spacing measured from the pattern corresponds to the d-spacing of the {111} planes of fcc Ag. (*d*) The size PDF and cumulative distribution function (CDF) of the AgSi NPs. (*e*) An STEM image showing the position of Ag cores (brighter contrast) within the Si shell.

achieved by increasing the sputtering power supplied to the Ag target from 4.7 W to 6.7 W (~30%).

3.3.3 The morphology, structure and composition of AgSi HNPs

A high magnification TEM image shows the multicore–shell morphology of the resulting AgSi NPs (figure 3.6(*a*)). The PDF, calculated over 100 NPs, shows that approximately 60% of the NPs are 10–20 nm in size and more than 25% are 20–30 nm in size (figure 3.6(*d*)). It is worth noting that no size selection was performed during the NP deposition. HRTEM images (figure 3.6(*b*)), in combination with the corresponding FFT pattern (figure 3.6(*c*)), show the crystalline structure of the cores. The d-spacing in the region indicated in figure 3.6(*b*) was determined to be 0.23 nm, corresponding to the d-spacing of the {111} planes of fcc Ag [15].

Figure 3.6(*e*) shows an STEM image of an individual NP in which one can observe the difference in contrast (due to the atomic numbers of Ag and Si) within the NP, indicating the positions of Ag cores (brighter contrast).

Complementary EELS analysis was performed to obtain further information on the chemical composition of AgSi multicore–shell NPs. From the EELS spectrum (figure 3.7(*b*)) corresponding to area 1 as indicated in the STEM image (figure 3.7(*a*)),

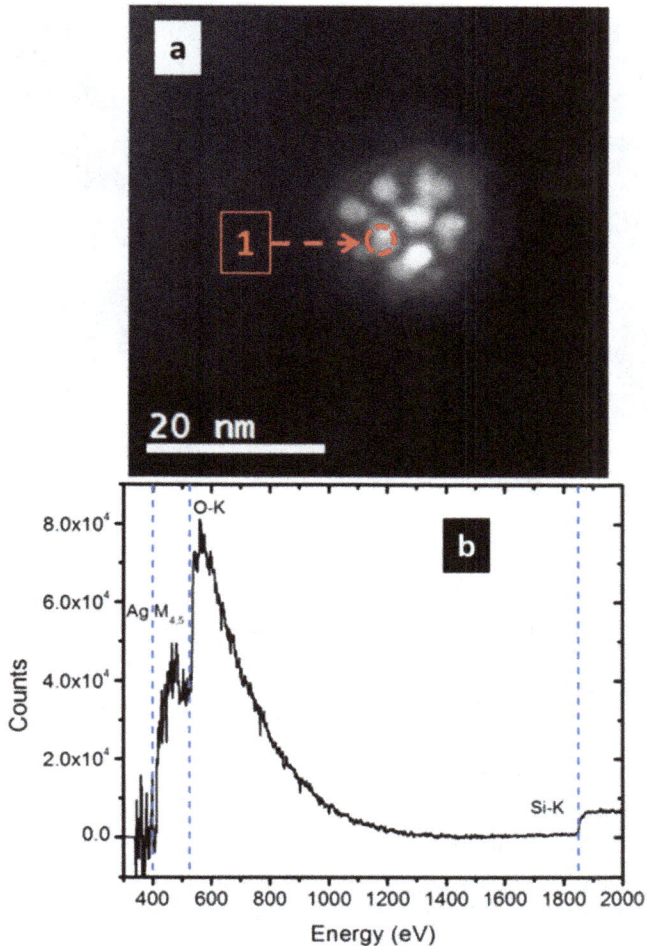

Figure 3.7. (*a*) An STEM image of a multicore–shell AgSi NP. (*b*) The EELS spectrum obtained from region 1 as indicated in (*a*), showing the Ag–$M_{4,5}$ (401 eV), O–K (532 eV) and Si–K (1860 eV) edges.

it is possible to see the Ag–$M_{4,5}$ (401 eV) and O–K (532 eV) edges, as well as the Si–K edge at higher energy (1860 eV). The high oxygen content is attributed to the oxidation of the Si shell when the sample was exposed to the atmosphere.

In the following experiment, the power on the Ag target was increased from 4.7 W to 6.7 W while the other parameters were kept constant as described in section 3.3.2. The resulting AgSi NPs clearly show a mixture of two types of morphologies, namely core–shell, core–satellite and core–shell–satellite structures (figure 3.8(*a*)). From the bright-field high magnification TEM image (figure 3.8(*c*)), it can be seen that the core–satellite NPs have a near-spherical core with a number of smaller clusters of other material deposited on their surfaces. From the high magnification STEM image of a representative core–satellite NP (figure 3.8(*b*)), it is obvious that

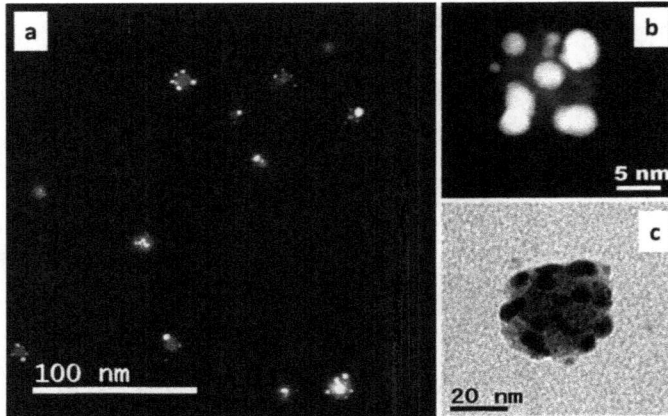

Figure 3.8. (*a*) An STEM image showing the core–satellite, multicore–shell and core–shell–satellite morphologies of the NPs. (*b*), (*c*) High magnification STEM and TEM images, respectively, of a representative core–satellite NP.

the small clusters deposited on the core are Ag-rich clusters (higher contrast = higher atomic number). The observed satellite morphology suggests that, due to the higher content of Ag in the aggregation zone, Ag clusters deposit on the surface of the NPs while in flight in the aggregation zone towards the deposition chamber.

3.3.4 The formation mechanism

This section suggests a simple formation mechanism for AgSi based on the miscibilities and surface energies of Ag and Si. Because of its high sputtering yield (~1.20 under our experimental conditions [9]), when an Ag target is being sputtered, electrons and ions are highly concentrated in the plasma region, inducing a high-density plasma near the Ag target surface; the nucleation and growth of the Ag NPs is then completed before leaving the plasma zone. In the aggregation zone, the density of inert gas is sufficient to quench their high energy. As a result, the particles retain their high temperature phase and larger nanoclusters with a mono-crystalline structure are formed (figure 3.9, step 1).

In contrast, silicon has a low sputtering yield of ~0.29 [9]; during Si target sputtering, the plasma density on the target surface is low and the nucleation and growth of the NPs therefore occur during the flight of atoms into the aggregation zone. Under these conditions, Si atoms lose their energy rapidly and form small and amorphous Si nanoclusters.

Due to the low surface energy of amorphous silicon (1.05 J m^{-2}) [31], Si clusters eventually cover the surface of the Ag NPs. Assisted by the lower melting temperature associated with their amorphous structure, these clusters are able to pre-melt and wet the surface of the Ag NPs, resulting in core–shell structures with shells of quite uniform thicknesses. Finally, during their flight through the aggregation zone, the Ag–Si core–shell NPs collide and partially coalesce with each other,

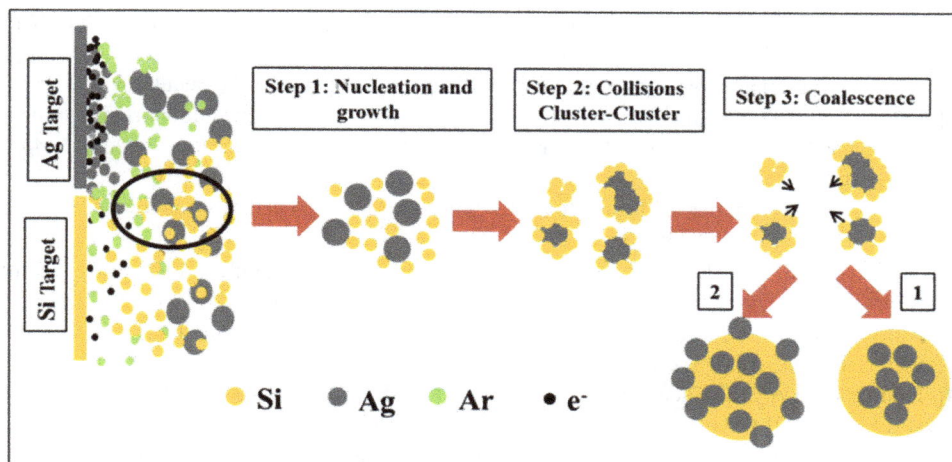

Figure 3.9. A diagram summarizing the suggested formation mechanism for the AgSi HNPs. Step 1 shows the nucleation and growth phase in the plasma zone near the surface of the sputtered targets. Step 2 depicts cluster–cluster collisions. Si NPs then cover the surface of the Ag NPs resulting in a core–shell structure. The resulting core–shell NPs coalesce with each other in-flight and form multicore–shell HNPs (step 3).

forming multicore–shell morphologies. The formation of a core–satellite morphology may be attributed to the higher amount of Ag due to the increased power on the Ag target.

References

[1] Krishnan G, Verheijen M A, Ten-Brink G H, Palasantas G and Kooi B J 2013 Tuning structural motifs and alloying of bulk immiscible Mo–Cu bimetallic nanoparticles by gas-phase synthesis *Nanoscale* **5** 5375–83

[2] Vernieres J, Benelmekki M, Kim J H, Grammatikopoulos P, Bobo J F, Diaz R E and Sowwan M 2014 Single-step gas phase synthesis of stable iron aluminide nanoparticles with soft magnetic properties *APL MATERIALS* **2** 116105

[3] Singh V, Cassidy C, Grammatikopoulos P, Djurabekova F, Nordlund K and Sowwan M 2014 Heterogeneous Gas-Phase Synthesis and Molecular Dynamics Modeling of Janus and Core–Satellite Si–Ag Nanoparticles *J. Phys. Chem.* C **118** 13869–75

[4] Llamosa Perez D, Espinosa A, Martinez L, Roman E, Ballesteros C, Mayoral A, Garcia-Hernandez M and Huttel Y 2013 Thermal diffusion at nanoscale: from CoAu alloy nanoparticles to Co@Au core–shell structure *J. Phys. Chem.* C **117** 3101–8

[5] Langlois C, Alloyeau D, Bouar Y L, Loiseau A, Oikawa T, Mottet C and Ricolleau C 2008 Growth and structural properties of CuAg and CoPt bimetallic nanoparticles *Faraday Discuss.* **138** 375–91

[6] Makino A, Hatanai T, Naitoh Y, Bitoh T, Inoue A and Masumoto T 1997 Applications of nanocrystalline soft magnetic materials Fe–M–B (M = Zr, Nb) alloys NANOPERM (R) *IEEE Trans. Mag.* **33** 3793–8

[7] Osaka T, Takai M, Hayashi K, Ohashi K, Saito M and Yamada K 1998 A soft magnetic CoNiFe film with high saturation magnetic flux density and low coercivity *Nature* **392** 796–8

[8] Margeat O, Ciuculescu D, Lecante P, Respaud M, Amiens C and Chaudret B 2007 NiFe nanoparticles: A soft magnetic material? *Small* **3** 451–8

[9] www.iap.tuwien.ac.at/www/surface/sputteryield a simple sputter yield calculator

[10] Quesnel E, Pauliac-Vaujour E and Muffato V 2010 Modeling metallic nanoparticle synthesis in a magnetron-based nanocluster source by gas condensation of a sputtered vapour *J. Appl. Phys.* **107** 054309

[11] Graupner H, Hammer L, Heinz K and Zehner D M 1997 Oxidation of low-index FeAl surfaces *Surf. Sci.* **380** 335–51

[12] Moulder J F, Stickle W F, Sobol P E and Bomben K D 1992 *Handbook of X-ray Photoelectron Spectroscopy,* ed J Chastain (Waltham, MA: Perkin Elmer)

[13] Yamashita T and Hayes P 2008 Analysis of XPS spectra of Fe^{2+} and Fe^{3+} ions in oxide materials *Appl. Surf. Sci.* **254** 2441–9

[14] Rodriguez G A C, Guillen G G, Palma M I M, Roy T K D, Hernandez A M G, Krishnan B and Shaji S 2014 Synthesis and characterization of Hercynite nanoparticles by pulsed laser ablation in liquid technique *Int. J. Appl. Ceram. Tec.* at press

[15] Benelmekki M, Vernieres J, Kim J H, Diaz R E, Grammatikopoulos P and Sowwan M 2015 On the Formation of Ternary Metallic–Dielectric Multicore–Shell Nanoparticles by Inert-Gas Condensation Method *Mater. Chem. Phys.* **151** 275–81

[16] Xu Y H and Wang J P 2008 Direct Gas-Phase Synthesis of heterostructured nanoparticles through phase separation and surface segregation *Adv. Mater.* **20** 994–9

[17] Yang J, Hu W, Tang J and Dai X 2013 The formation of Fe core Al shell and Fe shell Al core nanoparticles, a molecular dynamics simulation *Comp. Mater. Sci.* **74** 1604

[18] Yang J, Hu W and Tang J 2014 Effect of incident energy on the configuration of Fe–Al nanoparticles, a molecular dynamics simulation of impact deposition *RSC Adv.* **4** 2155–60

[19] Breuer J 2001 Statistical thermodynamics of ordered intermetallic compounds containing point defects *PhD Thesis* University of Stuttgart

[20] Alqudami A and Annapoorni S 2007 Fluorescence from metallic silver and iron nano-particles prepared by exploding wire technique *Plasmonics* **2** 5–13

[21] Zhai Y, Han L, Wang P, Li G, Ren W, Liu L, Wang E and Dong S 2011 Superparamagnetic Plasmonic Nanohybrids: Shape-Controlled Synthesis, TEM-Induced Structure Evolution, and Efficient Sunlight-Driven Inactivation of Bacteria *ACS Nano* **5** 8562–70

[22] Cade N I, Ritman-Meer T, Kwakwa K A and Richards D 2009 The plasmonic engineering of metal nanoparticles for enhanced fluorescence and Raman scattering *Nanotechnology* **20** 285201

[23] Nomoev A V and Bardakhanov S P 2012 Synthesis and structure of Ag–Si nanoparticles obtained by the electron-beam evaporation/condensation method *Tech. Phys. Lett.* **38** 375–8

[24] Sotiriou G A, Hirt A M, Lozach P Y, Teleki A, Krumeich F and Pratsinis S E 2011 Hybrid, Silica-Coated, Janus-Like Plasmonic-Magnetic Nanoparticles *Chem. Mater.* **23** 1985–92

[25] Aslan K, Wu M, Lakowicz J R and Geddes C D 2007 Fluorescent Core–Shell $Ag@SiO_2$ Nanocomposites for Metal-Enhanced Fluorescence and Single Nanoparticle Sensing Platforms *J. Am. Chem. Soc.* **129** 1524–5

[26] Chen K H, Pu Y C, Chang K D, Liang Y F, Liu C M, Yeh J W, Shih H C and Hsu Y J 2012 Ag-Nanoparticle-Decorated SiO_2 Nanospheres Exhibiting Remarkable Plasmon-Mediated Photocatalytic Properties *J. Phys. Chem.* C **116** 19039–45

[27] Mertens H 2007 Controlling plasmon-enhanced luminescence *PhD Thesis* Institute for Atomic and Molecular Physics, Amsterdam

[28] Olesinski R W, Gokhale A B and Abbaschlan G J 1989 The Ag–Si (Silver–Silicon) System *Bull. Alloy Phase Diagr.* **10** 635–40

[29] Bokhonov B and Korchagin M 2002 *In situ* Investigation of the Formation of Eutectic Alloys in the Systems Silicon–Silver and Silicon–Copper *J. Alloys Compd.* **335** 149–56

[30] Benelmekki M, Bohra M, Kim J H, Diaz R E, Vernieres J, Grammatikopoulos P and Sowwan M 2014 Facile Single-Step Synthesis of Ternary Multicore Magneto-Plasmonic Nanoparticles *Nanoscale* **6** 3532–5

[31] Hara S, Izumi S, Kumagai T and Sakai S 2005 Surface energy, stress and structure of well-relaxed amorphous silicon: A combination approach of *ab initio* and classical molecular dynamics *Surf. Sci.* **585** 17–24

Chapter 4

Design of ternary magneto-plasmonic nanoparticles

4.1 Introduction to magneto-plasmonic NPs

HNPs that contain both magnetic and optically active components have emerged as attractive candidates for advanced biomedical applications, such as multimodal bio-imaging, targeted drug delivery and magnetic hyperthermia. In particular, dumbbell-like NPs composed of magnetite (Fe_3O_4) and Au or Ag, are attractive because of their biocompatibility, plasmonic activity and magnetic properties [1–7]. A broad range of techniques in chemistry and physics has been explored for the preparation of such multicomponent NPs. Further details of the synthesis, properties and applications of dumbbell-like NPs can be found in several review articles [1, 8, 9].

In the last decade, physical vapour deposition techniques have been developed for production of hybrid inorganic NPs. However, only a few attempts have been made to produce heterogeneous, dumbbell-shaped multicomponents [10, 11] other than the widely reported core–shell nanostructures [12, 13]. In the context of this chapter, reports of the preparation of Fe-rich-core–Si-rich-shell NPs using double-glow discharge sources [14, 15] are worth noting. The behaviour of Ag–Fe magneto-plasmonic NPs (MPNPs) prepared by magnetron sputtering with subsequent in-flight annealing was reported in [16] and demonstrated the instability of these NPs under atmospheric conditions.

4.2 The deposition of FeAg@Si MPNPs

NPs were synthesized using the MS-IGC method as described in chapter 2. Three independent targets of Fe, Ag and Si were co-sputtered under controlled conditions (figures 2.2 and 2.3). One-inch diameter Si (99.999%), Ag (99.99%) and Fe (99.9%) targets were used as the ion sources for NP formation. The dc powers applied to the Si, Fe and Ag targets were 64 W, 10 W and 4.8 W, respectively. Ar flux in the aggregation zone was set at 80 sccm resulting in a working pressure of 3×10^{-1} mbar in the aggregation zone. The pressure at the deposition chamber was set at 7.5×10^{-4} mbar. The NPs were deposited on different substrates for TEM images: Si, Cu, graphite and carbon coated Cu grids. The working distance, from the target surfaces to the exit slit (figure 2.2(*a*)), was set at ~90 cm.

4.3 Characterization methods

TEM analysis was performed using a Cs-corrected-FEI-Titan 80–300 kV microscope operating at 300 kV (see figure 1.8). SEM images were collected using an FEI Quanta FEG 250 system. For XPS analysis, a Kratos Axis Ultra^{DLD} 39-306 electron spectrometer equipped with a monochromatic AlKα source operated at 300W was used (figure 1.9).

4.4 The morphology, structure and composition of FeAg@Si MPNPs

The size PDF and CDF of the resulting NPs were calculated from TEM images. As shown in figure 4.1(*b*), approximately 65% of the resulting MPNPs (hereafter MPNPs-1) were 10–20 nm in size and 20% were 20–30 nm in size. This size distribution can be correlated with the number of cores per NP, as reported in [17]. Figures 4.1(*c*)–(*e*) display high magnification TEM images of individual MPNPs showing multicore–shell morphology and irregular shapes. STEM images (figure 4.1(*f*)–(*h*)) show the differences in contrast (due to the atomic numbers of the constituents) within MPNPs-1, indicating the positions of Ag NPs (brighter contrast). HRTEM images of a representative four-core NP (figures 4.2(*a-1*) and (*a-2*)) show crystalline structure differences within the individual dumbbell-like shaped cores embedded in an amorphous shell. A STEM image of the same NP (figure 4.2(*b*)) shows the difference in contrast within the cores, indicating the positions of Ag NPs (brighter contrast).

For further information on the chemical composition of the cores, EELS analysis was performed on a two-core NP as shown in figure 4.3. Areas with brighter contrast in the dumbbell-like NP are Ag-rich. In fact, from the EELS spectrum corresponding to area d (figure 4.3(*d*)), it is possible to distinguish both the Ag–$M_{4,5}$ (401 eV) and O–K (532 eV) edges, as well as a small Fe–$L_{2,3}$ (710 eV) edge that can result from a small amount of Fe trapped in the Si shell. However, in the EELS spectrum corresponding to area c (figure 4.3(*c*)), only the Fe–$L_{2,3}$ edge and O–K edge are present (the oxygen arises from oxidation of the Si shell when exposed to air). In both cases, the EELS spectra show a Si edge at higher energy (1860 eV). From the HRTEM image (figure 4.3(*b*)) and associated FFTs (not shown), the Ag-rich region reveals the {111} Ag planes with an interplanar distance of 0.238 nm, while the Fe-rich and Si regions show an amorphous structure. Thus, this combination of techniques, i.e., bright field imaging, STEM imaging and EELS analysis, confirm the core–shell structure of the nanoclusters with a core of Fe–Ag dumbbell-like NPs covered with a Si shell.

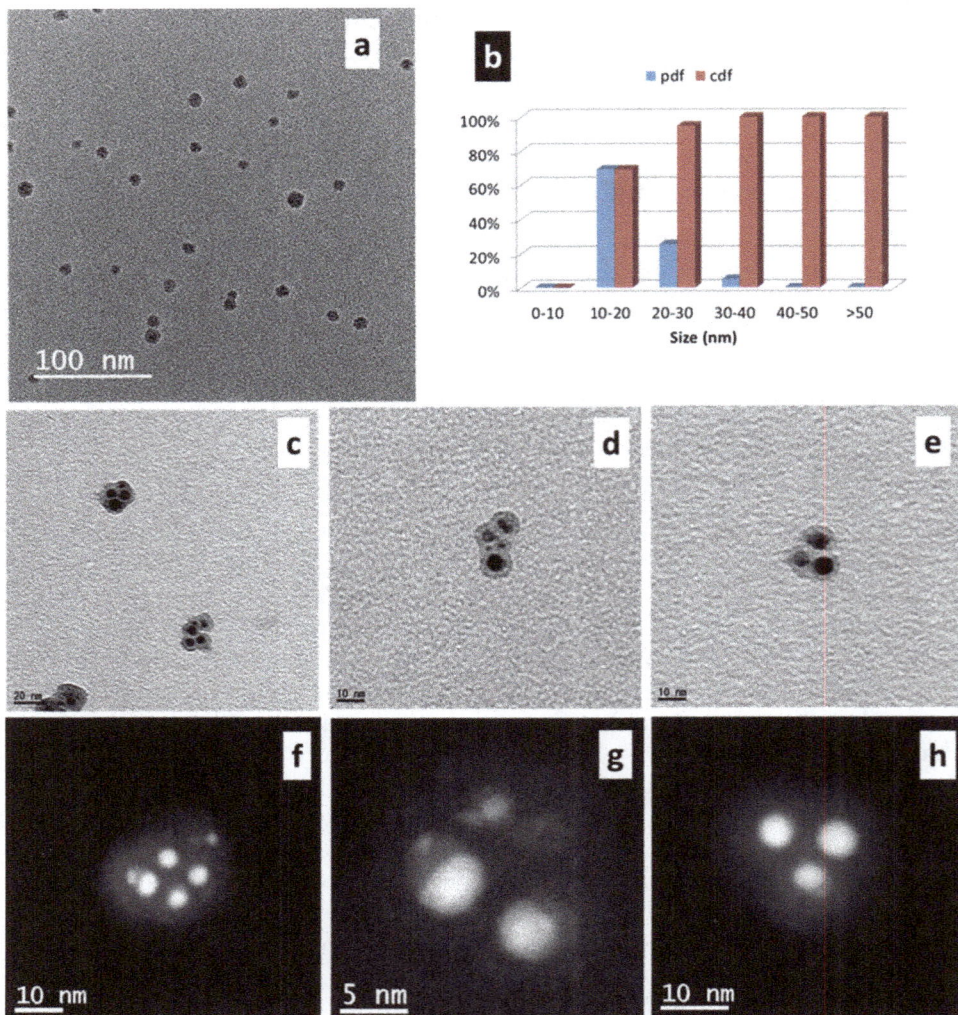

Figure 4.1. (*a*) A low magnification TEM image and (*b*) size PDF and CDF of MPNPs-1, calculated from TEM images. (*c*),(*d*),(*e*) Bright-field TEM images of individual FeAg@Si NPs showing the multicore–shell morphology and the irregular shapes of the NPs. (*f*),(*g*),(*h*) STEM images of FeAg@Si NPs showing the dumbbell-shaped cores. The brightness variations in (*f*), (*g*) and (*h*) indicate three different materials.

Another confirmation of the dumbbell-like structure of the cores can be achieved by scanning across one core as shown in figure 4.4. The line scan through the core, in the direction of the arrow as indicated in figure 4.4(*a*), shows an increasing Ag signal that reaches a peak at approximately 4 nm and then decreases to a minimum at approximately 7.5 nm. For the Fe signal, a defined minimum at ~4 nm was observed, followed by a distinct maximum at ~7.5 nm.

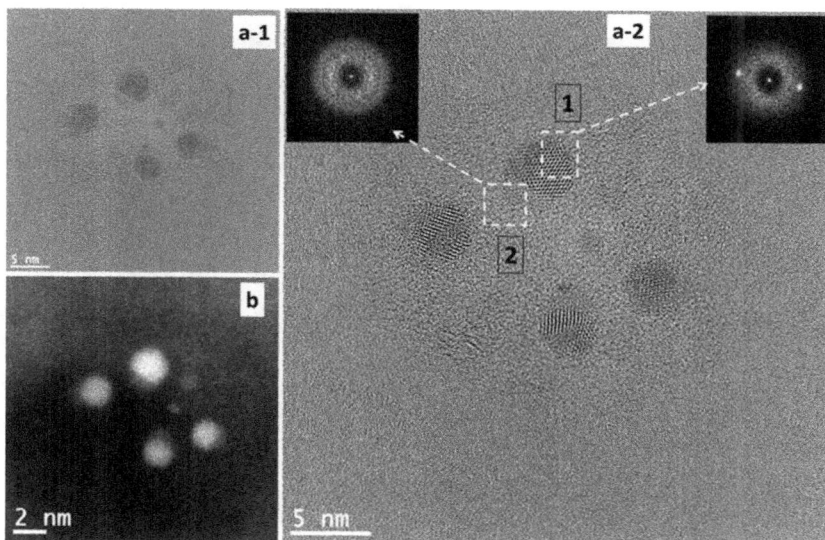

Figure 4.2. (*a-1*),(*a-2*) HRTEM images of the same four-core NP showing the differences in contrast within the dumbbell-like core and the Si amorphous shell. The insets in (*a-2*) show the FFT patterns corresponding to regions 1 and 2. The d-spacing measured from pattern 1 corresponds to the d-spacing of the {111} planes of fcc Ag. FFT pattern 2 demonstrates the amorphous structure of the Fe. (*b*) An STEM image of the same NP showing the brightness variation within the cores indicating two different materials.

4.5 The morphology and size tuning of MPNPs

4.5.1 The effects of the pressure differential between the aggregation zone and the deposition chamber

Increasing the pressure in the deposition chamber from 7.5×10^{-4} mbar to 3.9×10^{-3} resulted in the formation of MPNPs termed MPNPs-2 (figure 4.5(*a*)). In comparison with MPNPs-1, only 50% of MPNPs-2 are below 30 nm in size, 20% are 30–50 nm and 30% exceeded 50 nm (figure 4.5(*b*)). These results demonstrate that by reducing the pressure differential between the aggregation zone and the deposition chamber, the residence time of the NPs in the aggregation zone increases and the coalescence process is enhanced, leading to the formation of larger NPs with a higher number of cores and thicker Si shells (figure 4.5(*c*), (*d*) and (*e*)) [18]. EELS profiles were intended to determine the position of each component within the NPs; however, even for the smallest MPNPs-2, due to the thickness of the Si shell and the larger sizes of the NPs, it was difficult to determine whether the cores had a core–shell structure or a dumbbell-like shape (figure 4.6).

4.5.2 The effects of discharge powers on the Ag target

In this experiment, the power on the Ag target was increased from 4.8 W to 6.7 W while the other parameters were kept constant as described in section 4.2. The resulting MPNPs (hereafter MPNPs-3) clearly have a mixture of two types of

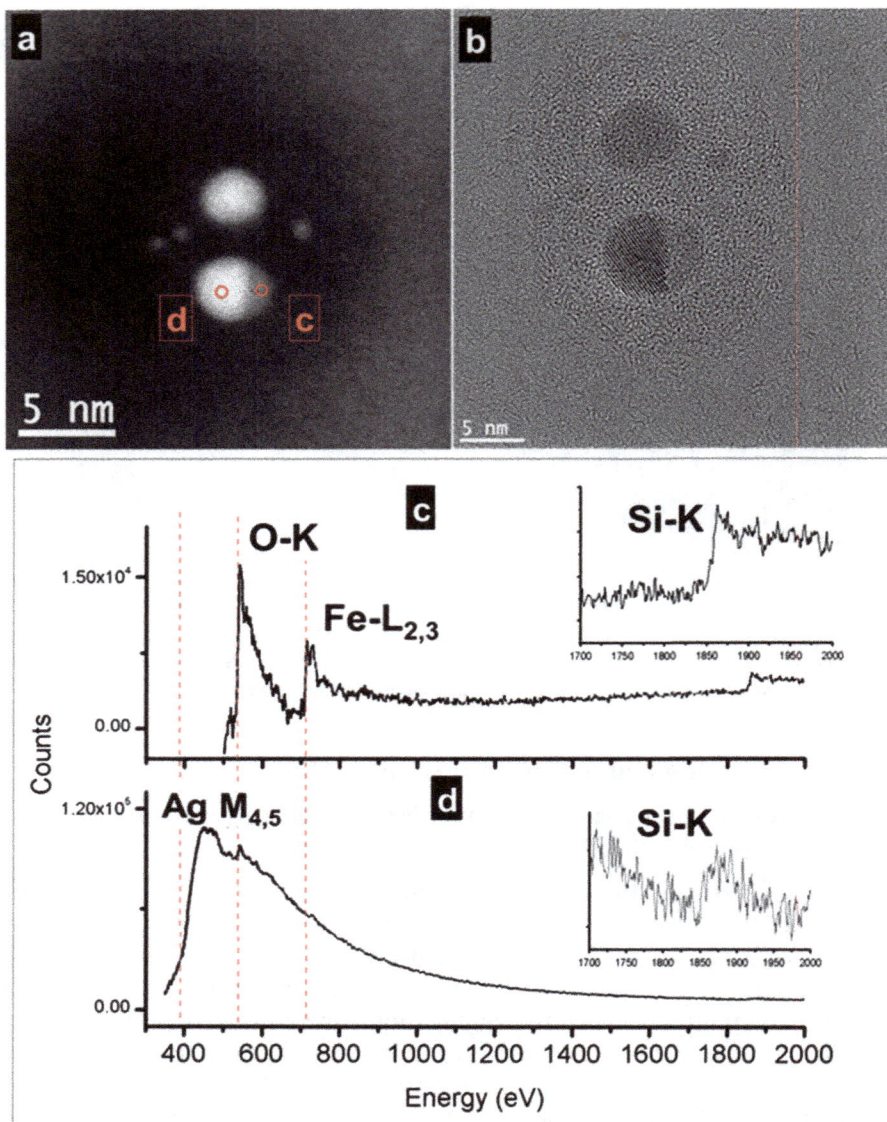

Figure 4.3. (*a*) STEM and (*b*) HRTEM images of the same NP. (*c*),(*d*) EELS spectra showing that the cores are composed of Fe and Ag domains in a dumbbell-like shape and that the shell is composed of Si. Reproduced from [17] by permission of The Royal Society of Chemistry.

structures, namely core–shell and core–satellite structures (figure 4.7(*a*) and (*b*)). The core–shell NPs have irregular shapes with well-defined cores of one material encapsulated by another material. The core–satellite NPs have a roughly spherical core with a number of smaller clusters of other material deposited on their surfaces. The observed satellite morphology suggests that, because of the higher content of Ag

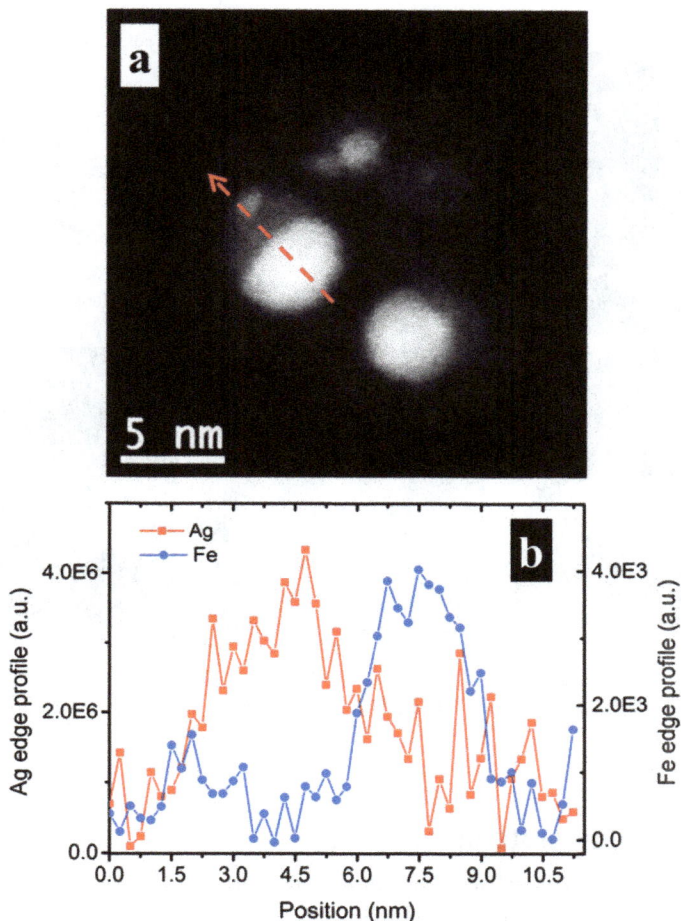

Figure 4.4. (a) An STEM image and (b) its corresponding EELS line scan showing the dumbbell-like structure of the Ag–Fe core. The red arrow in the STEM image represents the path where the beam was scanned.

in the aggregation zone, Ag clusters deposit on the surface of the formed NPs while in flight in the aggregation zone. Size distributions were obtained from TEM images. 80% of the NPs were 10–20 nm in size, clearly smaller than MPNPs-1 (figure 4.7(e)).

EELS line scans were performed across the particle in the direction of the arrow, as shown in figure 4.8. The Ag and Fe intensity profiles both show maxima at ~4.5 nm and ~9 nm positions, indicating a core–shell structure in both core NPs. The narrow shoulders of the Fe signal, appearing at ~7.5 and ~10.5 nm, indicate depletion of Fe during the growth process. These results suggest that by increasing the power on the Ag target, the Ag NPs formed have sufficient energy to mix with iron and segregate to form Fe–Ag core–shell NPs [11].

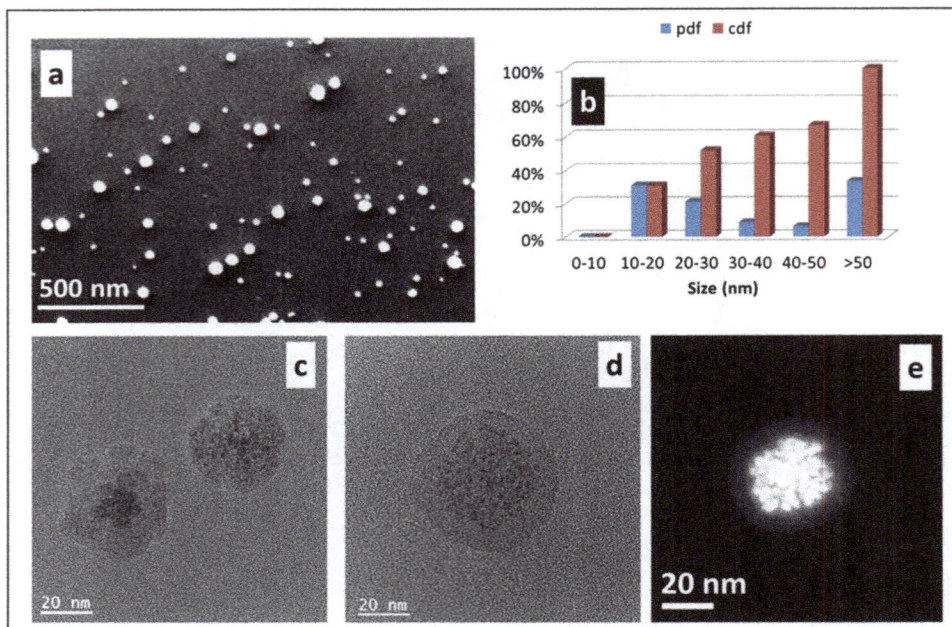

Figure 4.5. (*a*) An SEM image, the (*b*) size PDF and CDF of MPNPs-2, calculated from the SEM image. (*c*),(*d*) TEM images of different NPs and (*e*) an STEM image showing the multicore–shell morphology of the resulting NPs.

4.6 The oxidation state of MPNPs

XPS measurements were performed after exposure of the samples to the atmosphere. Fe $2p_{3/2}$ fitting peaks suggest that the iron cores in MPNPs-1 were dominated by the Fe^{3+} state (711.12 eV) (figure 4.9(*b*)). A first approximation of the relative contents of Fe, Ag and Si was calculated from Ag 3d, Fe 2p and Si 2p XPS core level peak areas (figure 4.9). The weight percentages (wt%) obtained for Fe, Ag and Si were ~33.82%, ~48.66% and ~17.51%, respectively. However, it is important to note that because of the surface oxidation effects, this calculation is not accurate, and thus is simply a rough approximation of the composition of the MPNPS-1 sample.

In contrast, in MPNPs-2 and MPNPs-3 the iron was present in both metallic (707.12 eV) and oxidized (711.33 eV) states [19, 20] (figure 4.10(*d*) and (*f*)). For Si 2p fitting peaks (figure 4.10(*a*), (*c*) and (*e*)), the atomic ratios between the Si2p@Si (99.11eV) and Si2p@SiOx (102.10 eV) were ~20% for MPNPs-1 and ~43% for MPNPs-2. In the case of MPNPs-3, the Si shell was nearly fully oxidized to SiO_x. These results suggest that in the case of MPNPs-2, because the Si shell was thicker than that of MPNPs-1 and MPNPs-3, a gradual oxidation occurred from the surface of the Si shell to the interface between the Si and the FeAg cores, providing better protection for the Fe cores against oxidation. In the case of MPNPs-3, despite the

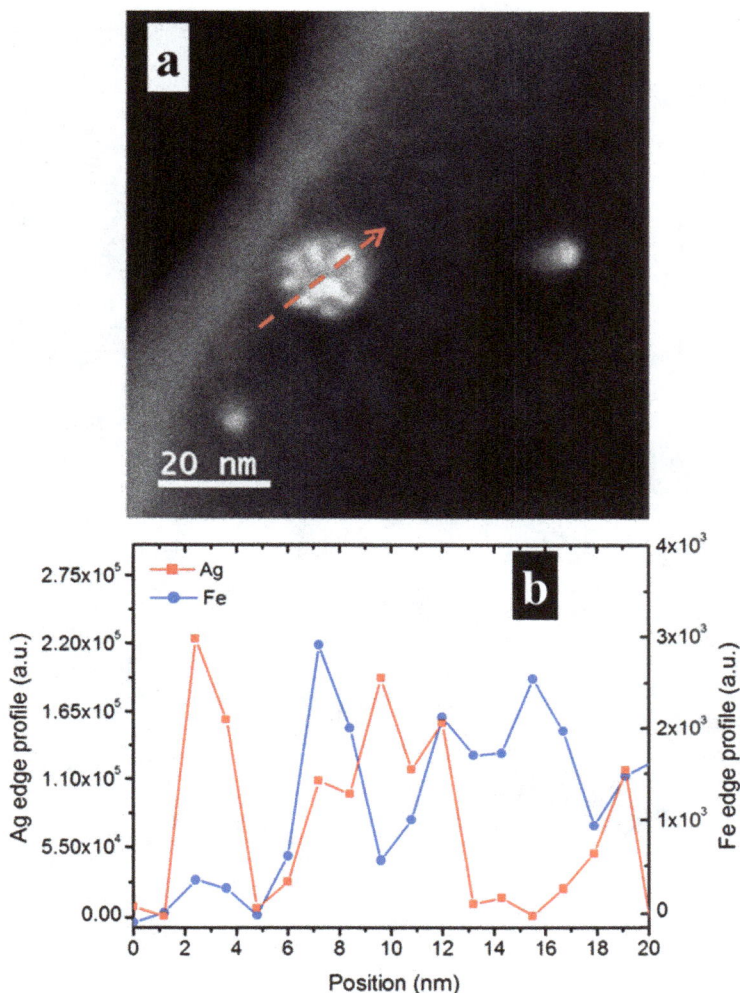

Figure 4.6. (*a*) An STEM image and (*b*) its corresponding EELS line scan of a NP selected from the MPNPs-2 sample. From these profiles, it is difficult to confirm the morphology of the cores. The red arrow in the STEM image represents the path of the scanning beam.

full oxidation of the Si shell, Fe was present in both its metallic and oxidized states. This behaviour was in agreement with EELS profiles, confirming that the iron was protected by the Ag shell (figure 4.8).

Regarding magnetic properties, the MPNPs exhibit a typical ferromagnetic behaviour with a coercivity of ~102 Oe at 300 K. These coercivities in all the samples can be attributed to the anisotropy at the FeAg interface. In addition, the UV–visible spectra corresponding to the MPNPs show an enhanced, red shifted, light absorption band due to the strong near-field coupling between the Ag cores and the Si shell. For more details on magnetic and optical properties see [17, 18, 25].

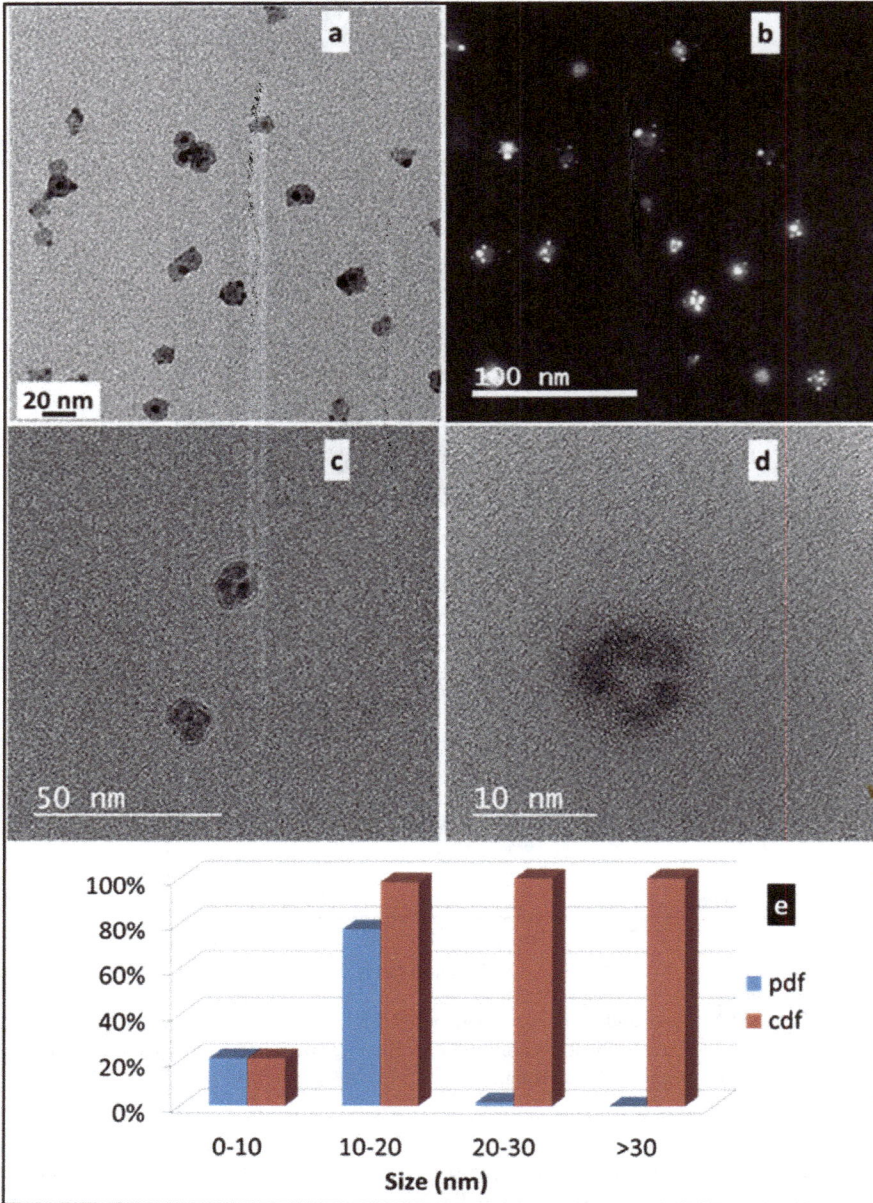

Figure 4.7. Low magnification TEM (*a*) and STEM (*b*) images of MPNPs-3 showing both core–shell and satellite morphologies (*c*) High magnification TEM (*d*) HRTEM images, respectively, showing the proximity of cores inside a Si shell and (*e*) The PDF and CDF of MPNPs-3 calculated from TEM images.

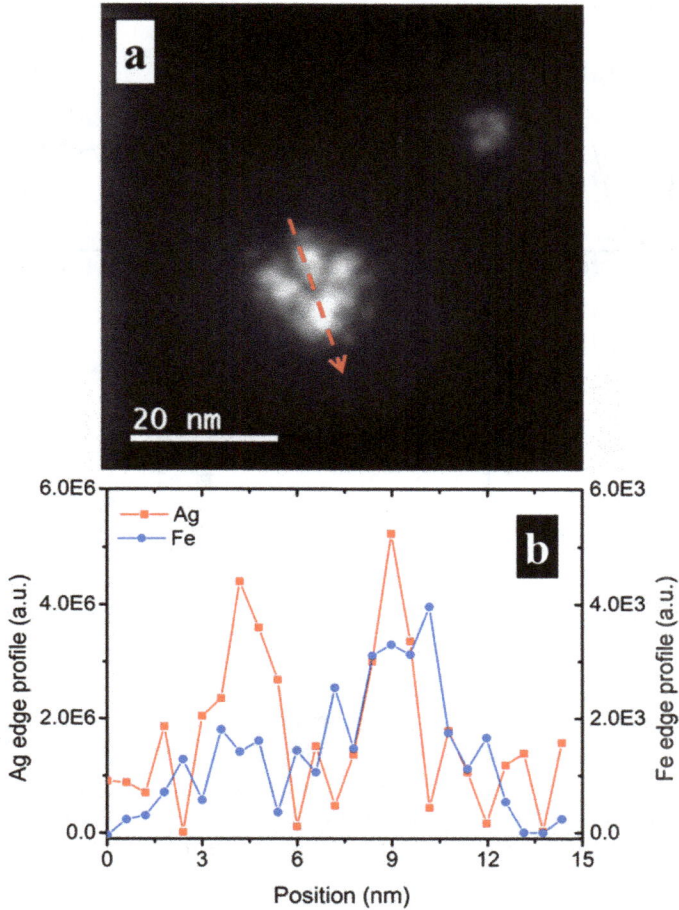

Figure 4.8. (*a*) An STEM image and (*b*) its corresponding EELS line scan showing the core–shell morphology of the Ag–Fe core. The red arrow in the STEM image represents the path where the beam was scanned.

4.7 The formation mechanism

Based on the fundamentals of magnetron sputtering, when the targets of Fe, Ag and Si are simultaneously sputtered, the plasma density plays a major role in the nucleation, growth and crystallization of the NPs. As previously explained (section 3.3.4), under the current experimental conditions, large nanoclusters of Ag with a mono-crystalline structure are formed. This is a result of the combination of two mechanisms: 1) the high-density plasma near the surface of the Ag target and 2) the sufficiency of the inert gas to quench the high energy of the formed Ag NPs towards the aggregation zone (figure 4.11, step 1).

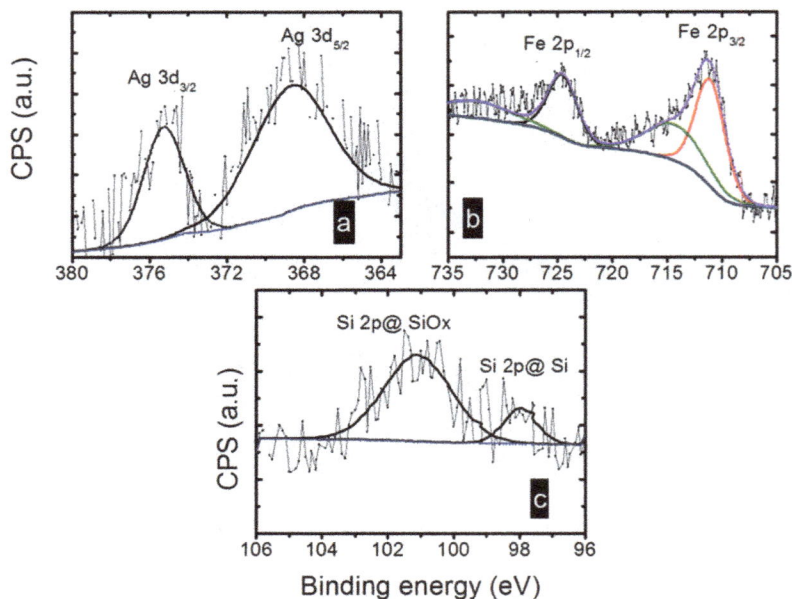

Figure 4.9. XPS spectra corresponding to (*a*) Ag 3d, (*b*) Fe 2p and (*c*) Si 2p core levels of the MPNPs-1 sample.

In contrast, for Fe NPs, the estimated sputtering yield is ~0.45 [23]. Moreover, the Fe target provides a path for the magnetic flux generated by the permanent magnet of the cathode (figure 2.4). Flux passes through the Fe target (figure 2.4(*c*)) and only a small portion escapes and reaches the surface, contributing to the generation of the magnetron plasma. Since the density and the distribution of the plasma are strongly correlated with the magnetic field strength and magnetic flux distribution [21, 22], during Fe sputtering only a low concentration of electrons near the surface is achieved. As a result, the ionization rate is lower than that of the Ag target, leading to a lower density of plasma. In this case, nucleation and growth of NPs occur during the flight of atoms into the aggregation zone. There, Fe atoms lose their energy rapidly and form small, amorphous Fe nanoclusters. The sputtering yield of Si is determined to be ~0.29 [23], inducing a low plasma density near the surface of the Si target. These experimental conditions lead to the formation of small and amorphous Si NPs.

Because the surfaces of the formed NPs are not oxidized, and the fraction of surface atoms per unit volume is very large, in the aggregation zone nanoclusters collide and coalesce by diffusion of the constituent atoms at their contact interfaces, creating larger NPs. Due to the large, positive free energy of Fe and Ag mixing, segregation of the elements forming core–shell or dumbbell-like structures would be expected [11, 16]. However, from our TEM study, the almost *exclusive* formation of dumbbell-like structures was observed in MPNPs-1. A plausible explanation for this behaviour is that when Fe and Ag NPs collide in the aggregation zone, their energy is not sufficient to induce complete coalescence, so that only dumbbell-like structures

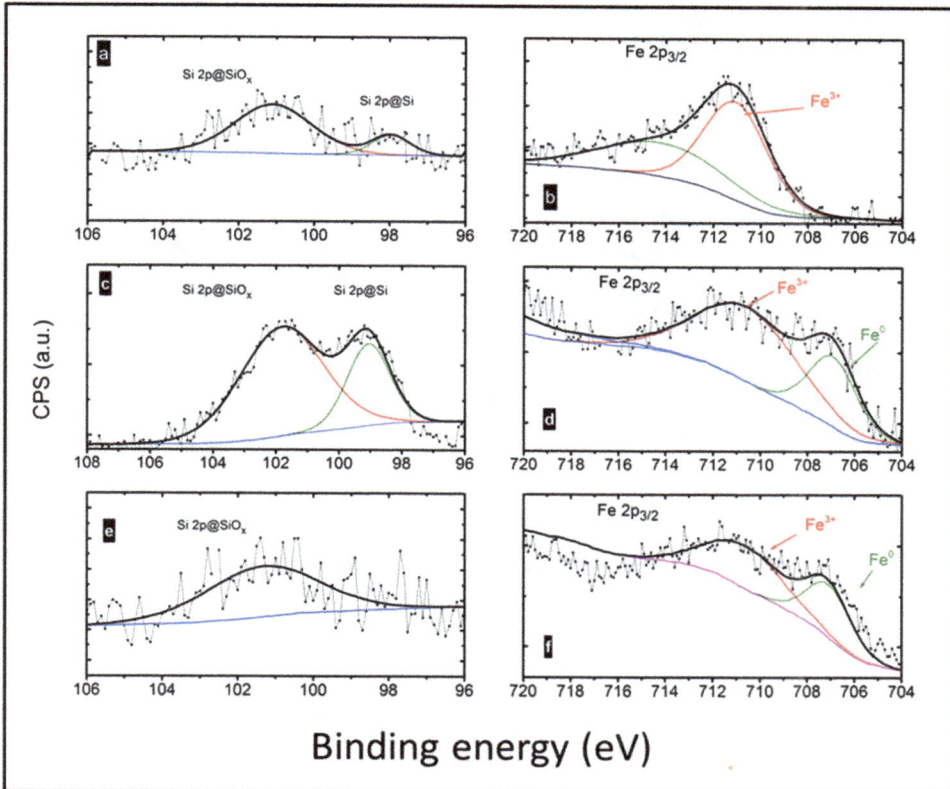

Figure 4.10. XPS spectra corresponding to Si 2p and Fe 2p core levels from MPNPs-1 (*a*),(*b*), MPNPs-2 (*c*),(*d*) and MPNPs-3 (*e*),(*f*).

can be formed. In the case of MPNPs-3, due to the higher current discharge applied to the Ag target, the pre-formed Ag NPs seem to retain enough energy to collide and mix with the Fe NPs, and then segregate to form an Fe–Ag core–shell structure (figure 4.11, step 2). Furthermore, due to the low surface energy of amorphous silicon (1.05 J m^{-2}) [24], Si clusters eventually cover the surface of the FeAg NPs. In addition, the lower melting temperature associated with the Si amorphous structure allows these clusters to pre-melt and wet the surface of the FeAg NPs, resulting in core–shell structures with shells of quite uniform thicknesses. Finally, during their flight through the aggregation zone, the FeAg–Si core–shell NPs collide and partially coalesce with each other, forming multicore MPNPs.

When the pressure differential between the aggregation zone and the main chamber is reduced, the residence time of the MPNPs in the aggregation zone is increased, increasing the probability of coalescence, resulting in larger NPs with a higher number of cores and thicker Si shells.

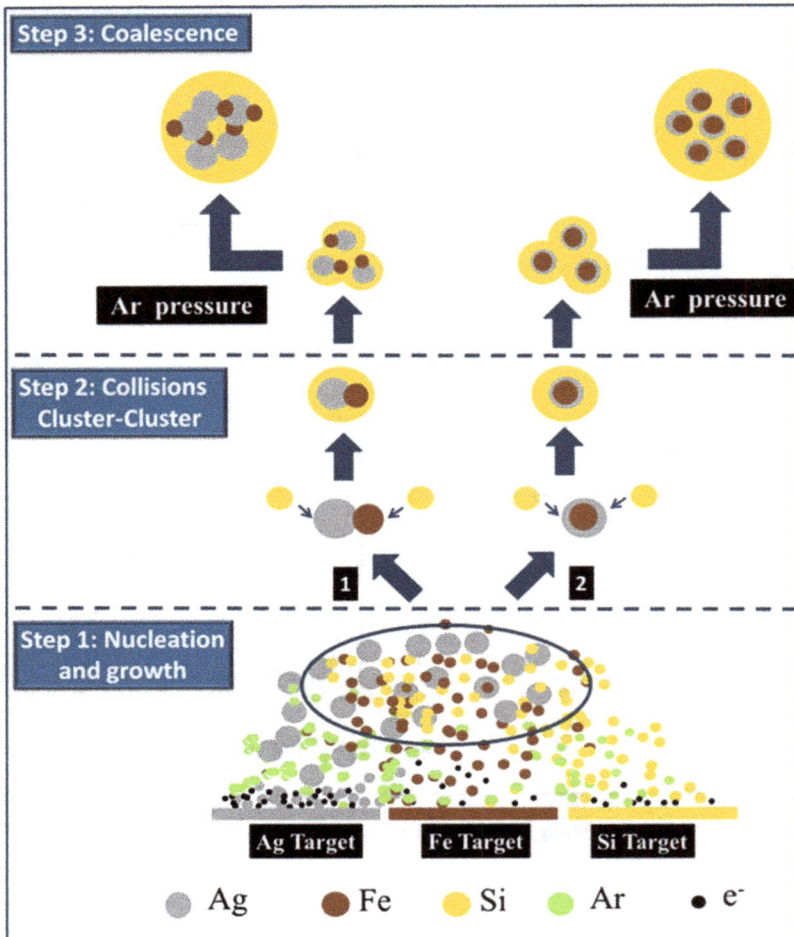

Figure 4.11. A diagram summarizing the suggested formation mechanism for of FeAg@Si MPNPs. Step 1 shows the nucleation and growth phase in the plasma zone near the surface of the sputtered targets. Step 2-1 depicts cluster–cluster collisions. Fe and Ag NPs juxtapose, forming dumbbell-like shaped NPs. In step 2-2 Ag and Fe NPs mix and then segregate in core–shell structures. Then Si nanoclusters cover the surface of the FeAg NPs resulting in a core–shell structure. The resulting core–shell NPs coalesce in-flight forming multicore–shell NPs (step 3).

References

[1] Wang C, Xu C, Zeng H and Sun S 2009 Recent Progress in Syntheses and Applications of Dumbbell-like Nanoparticles *Adv. Mater.* **21** 304552

[2] Parsina I, DiPaola C and Baletto F 2012 A novel structural motif for free CoPt nanoalloys *Nanoscale* **4** 1160–6

[3] Yu H, Chen M, Rice P M, Wang S X, White R L and Sun S 2005 Dumbbell-like bifunctional Au–Fe3O4 nanoparticles *Nano Lett.* **5** 379–82

[4] Gu H, Yang Z, Gao J, Chang C K and Xu B 2005 Heterodimers of nanoparticles: formation at a liquid–liquid interface and particle-specific surface modification by functional molecules *J. Am. Chem. Soc.* **127** 34–35

[5] Zhang L, Dou Y H and Gu H C 2006 Synthesis of Ag–Fe$_3$O$_4$ heterodimeric nanoparticles *J. Colloid Interface Sci.* **297** 660–4

[6] Lu Y, Xiong H, Jiang X and Xia Y 2003 Asymmetric Dimers Can Be Formed by Dewetting Half-Shells of Gold Deposited on the Surfaces of Spherical Oxide Colloids *J. Am. Chem. Soc.* **125** 12724–5

[7] Meng X, Seton H C, Lu L T, Prior I A, Thanh N T K and Song B 2011 Magnetic CoPt nanoparticles as MRI contrast agent for transplanted neural stem cells detection *Nanoscale* **3** 977–84

[8] Zeng H and Sun S 2008 Syntheses, Properties, and Potential Applications of Multicomponent Magnetic Nanoparticles *Adv. Funct. Mater.* **18** 391–400

[9] Perro A, Reculusa S, Ravaine S, Bourgeat-Lami E and Duguet E 2005 Design and synthesis of Janus nanoparticles *J. Mater. Chem.* **15** 3745–60

[10] Bai J, Xu Y H, Thomas J and Wang J P 2007 (FeCo)3Si–SiOx core–shell nanoparticles fabricated in the gas phase *Nanotechnology* **18** 065701

[11] Xu Y H and Wang J P 2008 Direct Gas-Phase Synthesis of heterostructured nanoparticles through phase separation and surface segregation *Adv. Mater.* **20** 994–9

[12] Skumryev V, Stoyanov S, Zhang Y, Hadjipanayis G, Givord D and Nogués J 2003 Beating the superparamagnetic limit with exchange bias *Nature* **423** 850–3

[13] Kaur M, McCloy J S, Jiang W, Yao Q and Qiang Y 2012 Size Dependence of inter-and intracluster interactions in core–shell iron–iron oxide nanoclusters *J. Phys. Chem.* C **116** 12875–85

[14] Sumiyama K, Katoh R, Kadowaki S and Hihara T 2010 Fe–Si core/Si-shell clusters prepared by double glow discharge sources *J. Nanopart. Res.* **12** 2589–96

[15] Tanaka N, Sumiyama K, Katoh R, Hihara T, Sato K, Konno T J and Mibu K 2010 Core-Shell Formation and Juxtaposition in Fe and Si Hybrid Clusters Prepared by Controlling the Collision Stages *Mater. Trans.* **51** 1990–6

[16] Elsukova A, Li Z A, Möller C, Spasova M, Acet M, Farle M, Kawasaki M, Ercius P and Duden T 2011 Structure, morphology, and aging of Ag–Fe dumbbell nanoparticles *Phys. Status Solidi* A **208** 2437–42

[17] Benelmekki M, Bohra M, Kim J H, Diaz R E, Vernieres J, Grammatikopoulos P and Sowwan M 2014 Facile Single-Step Synthesis of Ternary Multicore Magneto-Plasmonic Nanoparticles *Nanoscale* **6** 3532–5

[18] Benelmekki M, Vernieres J, Kim J H, Diaz R E, Grammatikopoulos P and Sowwan M 2015 On the Formation of Ternary Metallic–Dielectric Multicore–Shell Nanoparticles by Inert-Gas Condensation Method *Mater. Chem. Phys.* **151** 275–81

[19] Yamashita T and Hayes P 2008 Analysis of XPS spectra of Fe^{2+} and Fe^{3+} ions in oxide materials *Appl. Surf. Sci.* **254** 2441–9

[20] Moulder J F, Stickle W F, Sobol P E and Bomben K D 1992 *Handbook of X-ray Photoelectron Spectroscopy* Ed Jill Chastain (Waltham, MA: Perkin Elmer)

[21] Nanbu K and Kondo S 1997 Analysis of Three-Dimensional dc Magnetron Discharge by the Particle-in-Cell/Monte Carlo Method *Jpn. J. Appl. Phys.* **36** 4808–14

[22] Ido S, Kashiwawagi M and Takahashi M 1999 Computational Studies of Plasma Generation and Control in a Magnetron Sputtering System *Jpn. J. Appl. Phys.* **38** 4450–4

[23] www.iap.tuwien.ac.at/www/surface/sputteryield a simple sputter yield calculator

[24] Hara S, Izumi S, Kumagai T and Sakai S 2005 Surface energy, stress and structure of well-relaxed amorphous silicon: A combination approach of *ab initio* and classical molecular dynamics *Surf. Sci.* **585** 17–24

[25] Benelmekki M and Sowwan M 2015 *Handbook of Nano-Ceramic and Nano-Composite Coatings and Materials* ed H Makhlouf and D Scharnweber (Amsterdam: Elsevier) at press, ch 20

Chapter 5

Summary and future outlook

Substantial progress has been made in recent years in designing and synthesizing multicomponent NPs that combine two or more component materials into one nanostructure. Such complex structures have the potential to combine magnetic, plasmonic, semiconducting and other physical or chemical properties into a single object, allowing an enhanced functionality. The first efforts began with core–shell HNPs and then culminated in morphologies that deviate significantly from core–shell structures, such as heterodimers, or NPs decorated with several domains. In all cases, the enhanced and often new functionality resulting from the synergetic combination of parent properties allows the expansion of the already wide applicability of HNPs.

HNPs in which spherical, rod or branched domains are connected in a controlled morphology are promising for a number of potential uses ranging from nanoelectronics to biomedical, photovoltaic and catalytic applications. For example: 1) magneto–semiconductor NPs are very interesting as bimodal tags offering the ability to visualize labelled cells using both magnetic resonance and fluorescence imaging techniques, while an external magnetic field may be employed for the directed assembly of such materials; 2) MPNPs have been proposed for a large number of dual magneto-optical applications, theranostics, multimodal imaging, cell-sorting and biomagnetic-separation; 3) metal–semiconductor heterodimers are ideal candidates for solar energy harvesting applications and, in particular, photocatalysis where the semiconductor part of the NPs can be tailored for optimal light absorption, while the metal part facilitates charge separation.

Despite the wide variety of synthetic strategies that have been reported for the production of such advanced complex nanostructures, the development of new synthetic strategies that lead to finer morphological control of NPs is still a challenging task. Beyond the control of NP morphology, one of the most important open questions is to understand how synthetic conditions can be controlled to combine several materials with different structural and physical properties.

Within this context, the MS-IGC method is becoming one of the most widespread methods for the production of supported HNPs. This method allows accurate control of the morphology, chemical composition and size distribution of the NPs. In addition, it offers the advantage of producing phase-separated NPs with free-residual contamination both at the interface between the different components and on the surface.

This book presents three examples of HNPs produced using a modified MS-IGC method. The first example addresses the synthesis and characterization of intermetallic FeAl NPs whose components are partially bulk-miscible. Working under a vacuum allows the advantage of better control of the NP oxidation process. The resulting NPs consist of a nanoalloy of an iron aluminide crystalline core (DO3 phase) encapsulated in an alumina shell that reduces inter-particle interactions and prevents further oxidation and segregation of the FeAl core. The second example addresses the synthesis of metallic–dielectric Ag–Si NPs, whose materials are bulk-immiscible. In this case, an appropriate tuning of the sputtering power ratio of the Si and Ag independent targets allow the production of NPs with distinctive multicore–shell and core–satellite morphologies.

The third example consists of ternary NPs composed of magneto-plasmonic FeAg cores encapsulated within a Si shell. The size of the NPs and the shape of the cores in these NPs are modulated by controlling the deposition parameters as follows:

- Adjusting the discharge current on each target allows control of the plasma density on the surface of each target and thus the size and crystallinity of the formed NPs.
- Reducing the differential pressure between the aggregation zone and the deposition chamber increases the residence time of the NPs inside the aggregation zone enhancing the coalescence process and, therefore, larger NPs with a higher number of cores are formed.
- The immiscibility of Fe and Ag and the low surface energy of the amorphous Si play important roles in the position of each material within the HNPs.

In addition, when these NPs are exposed to the atmosphere, the Si surface is oxidized to SiO_x, providing an easy path for their surface modification for a wide range of bioapplications.

However, to gain a better understanding of NP formation, further improvements in characterization methods for aggregation zone conditions are needed. Moreover, to make the scale-up of this method profitable, there is an urgent need to improve the yield and harvesting processes of the resulting HNPs.

www.ingramcontent.com/pod-product-compliance
Lightning Source LLC
Chambersburg PA
CBHW081554220326
41598CB00036B/6671